ROADS

DRIVING AMERICA'S
GREAT HIGHWAYS

Larry McMurtry

A TOUCHSTONE BOOK
PUBLISHED BY SIMON & SCHUSTER
NEW YORK   LONDON   TORONTO   SYDNEY   SINGAPORE

TOUCHSTONE
Rockefeller Center
1230 Avenue of the Americas
New York, NY 10020

The Library of Congress has cataloged the Simon & Schuster edition as follows:
McMurtry, Larry.
Roads : driving America's great highways / Larry McMurtry.
p. cm.
1. United States—Description and travel. 2. Roads—United States. 3. McMurtry,
Larry—Journeys—United States. 4. Automobile travel—United States. I. Title.
E169.04.M39   2000   00-027889
917.304'929—dc21
ISBN 0-684-86884-9
0-684-86885-7 (Pbk)

For Michael Korda

ROADS

I WANTED TO DRIVE the American roads at the century's end, to look at the country again, from border to border and beach to beach. My son, James, a touring musician who sees, from ground level, a great deal of America in the line of duty, says that when it isn't his turn to drive the van he likes to sit for long stretches, looking out the window. "There's just so much to see," he says, and he's right. There's just so much to see.

From earliest boyhood the American road has been part of my life—central to it, I would even say. The ranch house in which I spent my first seven years sits only a mile from highway 281, the long road that traverses the central plains, all the way from Manitoba to the Mexican border at McAllen, Texas. In winter I could hear the trucks crawling up 281 as I went to sleep. In summer I would sit on the front porch with my parents and grandparents, watching the lights of cars as they traveled up and down that road.

We were thoroughly landlocked. I had no river to float on, to wonder about. Highway 281 was my river, its hidden reaches a mystery and an enticement. I began my life beside it and I want to drift down the entire length of it before I end this book.

Other than curiosity, there's no particular reason for these travels—just the old desire to be on the move. My destination is also my route, my motive only an interest in having the nomad in me survive a little longer. I'm not attempting to take the national pulse, or even my own pulse. I doubt that I will be having folksy conversations with people I meet as I travel. Today, in fact, I drove 770 miles, from Duluth, Minnesota, to Wichita, Kansas, speaking only about twenty words: a thank-you at a Quik Stop south of Duluth, where I bought orange juice and doughnuts; a lunch order in Bethany, Missouri; and a request for a room once I got to Wichita. The development of credit-card gas pumps, microwaves, and express motels has eliminated the necessity for human contact along the interstates. It is now possible to drive coast to coast without speaking to a human being at all: you just slide your card, pump your gas, buy a couple of Hershey bars, perhaps heat up a burrito, and put the pedal back to the metal.

Some years ago Mr. William Least Heat Moon wrote an appealing travel book in which he described the travels he had made in America on the blue highways—that is, the small roads that wind like vines across the land: the local roads, not the interstates. In the same spirit the novelist Annie Proulx has been known to travel as much as two thousand miles around the American west using only the *dirt* roads. These are real achievements, but achievements I have no desire to emulate. I intend to travel mainly on the great roads, the interstates: my routes will be the 10, the 40, the 70, 80, and 90; or if I'm in the mood to go north-south, I will mostly use the 5, the 25, 35, 75. The 95 I intend to ignore. I will, from time to time, switch off the interstates onto smaller roads, but only if they provide useful connectives, or take me to interesting places that the great roads—whose aim is to move you, not educate you—don't yet go.

It may be that Annie Proulx and William Least Heat Moon are successfully—if a little masochistically—probing America's heartlands. I salute them, but that's not what I want to do. I merely want to roll along the great roads, the major migration routes that carry Americans long distances quickly, east-west or north-south. What I really want to do is look. Indeed, I would have called this book *Just Looking* if John Updike hadn't already used that title for a collection of his essays on art.

Three passions have dominated my more than sixty years of mostly happy life: books, women, and the road. As age approaches, the appetite for long drives may leave me, which is why I want to get rolling now.

As background I should say that I own three thousand travel books, have read them, and given them a certain amount of thought. I began to read travel books in my youth, out of an interest in the places traveled to, most of which were beyond my purse. Then I got interested in the travelers who had felt the need to go to these often inhospitable places; and finally, now, I've come to have a critic's interest in travel narrative and have turned again to the famous travel writers—Burton, Thesiger, Doughty, Robert Byron, Cherry-Garrard, and others—in the hope of seeing how travel books are made before I attempt to make one myself. Does the route dictate the method, or vice versa? And why do the worst journeys make the best books?

I have also read a fair amount about the great roads or routes of old, the famous caravan routes, particularly the Silk Road out of Asia and the spice and salt routes in Africa, mainly out of an interest in nomadism itself and in the desire humans seem to have to migrate, even though the routes of migration are hard. Trade has usually been the motive for travel on the routes, but the need to be on the move may be an impulse deeper than trade.

Though the interstate highways began to crisscross America in the late sixties, it was the seventies before they became the great roadways that they are now. Not far from these roads lie the remains of earlier, once heavily traveled routes: highway 1, up the east coast (often swollen into a multilane highway now), route 66, U.S. 40, and the rest. When I first started driving around America in the mid-fifties I drove, of necessity, on many of these roads— narrow, cracked, crowded two-lane affairs. I doubt that now I'll feel much interest in following what remains of these old, difficult roads. Route 66, the most famous of all, parallels for much of its route what is now I-40 (not to be confused with U.S. 40, the so-called National Road, which runs through the heartland well to the north of I-40). In Texas and New Mexico one can pick up route 66 memorabilia at almost every gas station and Quik Stop. Not many of the oldsters who drove route 66 in its heyday will be apt to wax nostalgic about it, for it was always a dangerous road, with much more traffic to carry than it could carry safely. Dead bodies in the bar ditch and smushed cars on wreckers were always common sights along old 66.

My aim in recording these journeys is simple: to describe the roads as I find them and supplement current impressions with memories of earlier travels along some of the same routes. I'm driving for pleasure, and will consider myself quite free to ignore roads I don't like (that is, the 95) or to switch roads if I come to a stretch that bores me. My method, to the extent that I have one, is modeled on rereading; I want to reread some of these roads as I might a book. I recently reread *War and Peace,* skipping all of the Freemasonry and most of the philosophy of history. In the same spirit I expect to skip large chunks of the 10, the 40, and so on. Conversely, just as there are passages in Tolstoy of which I never tire, there are stretches of road whose beauty I can never exhaust, an example being the wonderful stretch of rangeland south of Emporia, Kansas, on the 35, where dun prairies stretch away without interruption to very distant horizons, with not one tree to violate one's sight line. Here there are even bovine overpasses, bridges over the 35 that lead to no highway but just allow cattle to graze both sides of the road.

On the other hand there are parts of this same highway, the 35, that I never want to drive again, the principal one being the long stretch from Dallas to San Antonio—an old, crumbling interstate that passes through endlessly repetitive stretches of ugly urban sprawl.

The repetitiveness of that particular stretch of highway, with the same businesses clustered at almost every exit, tempts me to advance one modest thesis, a counterargument to the often expressed view that because of the chain businesses, America all looks the same. But it doesn't, and it won't, no matter how many McDonald's and Taco Bells cluster around the exits. There are, after all, McDonald's in both Moscow and Paris, but few would argue that Russia and France look the same. In America the light itself will always differ; the winter light on the Sault Ste. Marie, at the head of the 75, will never be like the light over the Everglades, at the bottom of that road. Eastern light is never as strong or as full as western light; a thousand McDonald's will not make Boston feel like Tucson. Cities and suburbs and freeway exits may collect the ugliness of consumer culture, but place itself cannot be homogenized. Place will always be distinct, and these notes will show, I hope, that America is still a country of immense diversity—the north ends of the roads I'm planning to drive will never be like their south ends, nor will the east be like the west.

I have only a few days each month in which I'm free to drive, and want to disclaim right now any intention to be comprehensive when it comes to American roads. I may, for instance, skip the east entirely. There was a time when, like the heroine of Joan Didion's novel *Play It As It Lays*, I took a certain interest in the differing character of America's urban freeways; I could, then, have discoursed at some length about how the San Diego Freeway differs from the Long Island Expressway. But the days when the swelling out of our great urban freeways excited me are long past. Now I'd rather be almost anyplace on the planet—always excepting Rock Springs, Wyoming—than on the Long Island Expressway at rush hour. People who complain about congestion on the Los Angeles freeways have mostly never been on the freeways of the eastern corridor at the wrong time of day—which, increasingly, is *any* time of day. To see the Culture of Congestion at its most intense, just go east of Cleveland, north of Richmond, Virginia, south of Maine. There are still lovely drives to be found in New York State, in Vermont, in Maine, but getting to them nowadays requires a passage through a Calcutta of freeways and I'm not sure I'll want to trap myself on those roads again.

In general I think my preference will be to drive from worse to better, which usually means going from the north to the south or from the east to the west. If I decide to investigate I-90, for example, I would probably start in Boston and head for Seattle, rather than vice versa. If it's I-75 that interests me, I'd probably go from northern Michigan to southern Florida. Generally I'd rather be heading south, toward warmth, or west, toward bigger skies and stronger light. It's always uplifting to me to watch the opening of the land and the widening of the skies as I drive west, out of the forested country. This feeling, I suspect, is testimony to how determinative one's primal geography is. I had a plains upbringing and something in me responds to the plains as to no other landscape.

As a reader of many travel books, I have been impressed by the extraordinary stamina of the real explorers, from Mungo Park to Wilfred Thesiger. In contrast, I hardly feel that my little spurts along the interstates deserve to be called travels at all. Ney Elias, the English official who surveyed so much of central Asia, made quite prodigious journeys lasting several months, while receiving little recognition for these splendid efforts from his government. In the course of writing his great study *Siberia and the Exile System,* George Kennan (the nineteenth-century traveler, not his relative the twentieth-century diplomat) bounced around Siberia for so long, in conditions of such great discomfort, that he permanently damaged his health. In the twentieth century a host of explorers and travelers to all points of the globe—the poles, the deserts, the jungle, the great peaks—managed to find an abundance of hardships against which to test themselves. Even within the last few decades, James Fenton, Bruce Chatwin, Redmond

O'Hanlon, Colin Thubron, Dervla Murphy, and others have habitually managed to get themselves to places where comforts were scarce, though all of them, in the stiff-upper-lip tradition of such predecessors as Peter Fleming, Ella Maillart, Evelyn Waugh, Robert Byron, and Graham Greene, are casual and even airy about the annoyances and dangers of their journeys.

I am not of this company. Hardship is not something I seek, or even accept. I cheerfully confess that if the Hotel du Cap in Cap d'Antibes were a chain, I'd stay there every night, though I can be tolerably content with the more modest comforts afforded by the Holiday Inns, of which I've now stayed in about two hundred. I don't think I would enjoy the Silk Road, particularly—if I remember correctly, S. J. Perelman didn't like it very much, and he was a man whose tastes accord fairly closely with my own.

But the vanity of suggesting that my sails along the great roads amount to "travel," in the accepted sense of that term, can be well illustrated by the absurd brevity of my first little trip. I wanted to drive the 35 south from its origins in Duluth, Minnesota, and so flew north and spent an evening in a comfortable hotel looking down on Lake Superior. I left Duluth the next morning at 6 A.M. and by noon the next day was back in my bookshop in Archer City, checking my mail. I had come eleven hundred miles south in a day and a half, and I didn't even have to drive after dark. I left Duluth a little before first light and was comfortably installed in a Marriott in Wichita, Kansas, 770 miles to the south, well before sunset—and this was in January, when the sun still sets reasonably early. On the 35 my 770 miles didn't even feel like a hard day's drive. I was in a sensible Buick, drove at only modestly illegal speeds, and could easily have gone farther, had there been any reason to. What made it possible to travel that distance in an easy day was the great road itself, a highway designed for just that type of travel. I never had to go more than one hundred yards off the highway for food, gasoline, or a rest room. I had to slow only slightly for the cities I passed through, those being Minneapolis–St. Paul and Kansas City. (Des Moines was bypassed so neatly that I didn't have to slow down at all.)

The question that might be asked is, if that's all there is to driving the great roads, then what's to write about?

To answer the question I would look for a moment at history. It is my belief that the true predecessors of the interstates were not the little roads mentioned earlier—U.S. 40, route 66, and the like. The great roads of nineteenth-century America were the rivers of the Americas: the Hudson, the St. Lawrence, the Delaware, the Susquehanna, the Monongahela, the Ohio, the Arkansas, the Columbia, the Red River of the North, the Rio Grande, and of course, the Missouri-Mississippi. Most of those were navigable for much of their length, but even the rivers on which navigation was chancy or impossible still provided paths through the wilderness. The Missouri took Lewis and Clark a long way on their journey to the western ocean. For each of these rivers, as for the other great rivers of the world—the Nile, the Yangtze, the Volga, the Amazon, the Danube, the Niger, the Ganges—there is an abundance of travel narratives, some of them by explorers who had struggled up the rivers seeking their source—think of the Nile and its literature—and others by travelers who merely floated down the rivers, seeing the sights.

What I want to do is treat the great roads as rivers, floating down this one, struggling up that one, writing about these river-roads as I find them, and now and then, perhaps, venturing a comment about the land beside the road. For the road, like a river, very often merely passes through long stretches of countryside, having little effect on the lives of people who live only a few miles from it. When I lay abed as a boy in our ranch house, listening to those trucks growl their way up highway 281, the sound of those motors came to seem as organic as the sounds of the various birds and animals who were apt to make noises in the night. The sounds of cars and trucks blended in naturally, it seemed to me then, with the yips of coyotes, the lowing of cows, the occasional bellow of a bull, the whinnying of our horses, the complaints of the roosters, the hoot of an owl, and so on. The sounds of the road were part of the complex symphony of country life. Yet we had little to do with that road. I never went farther up it than Wichita Falls, twenty-eight miles north, where there was a cattle auction. I seldom went south on it at all. The road was just there, as it is to millions of Americans who live beside roads great and little. The roads are just there, routes to migrate along, if it's time to migrate.

Of the many downriver books I have read my favorite is *Slowly down the Ganges*, by the English traveler Eric Newby, about a twelve-hundred-mile trip he made in 1961, with his wife, Wanda, down India's holy river, from Hardwar to the Hooghly at Calcutta. Eric Newby has written several good travel books, one of which, *A Short Walk in the Hindu Kush*, is on the whole funnier than *Slowly down the Ganges;* but the latter is richer, in part because the life along the great river—that is, the life of India—is so rich.

Slowly down the Ganges is a book I've read many times. Usually, when I travel, I take it with me as a talisman, in the Penguin Travel Library edition. It has a quality very seldom found in travel books, and that quality is wisdom. In the main the great travelers, male or female, tend to be obsessed people; only obsession would get them across the distances they cross, or carry them through the hardships they face in the deserts, in the jungles, on the ice. They seldom attain and could perhaps not really afford wisdom, since wisdom, in most cases, would have kept them from ever setting out. Thus their narratives, particularly the greatest—of Thesiger, Burton, Lawrence, Doughty—are apt to seem to the reader to be the cries of solitary souls: exciting cries sometimes, joyous sometimes, weary and despairing at times, but always the cries of essentially solitary men working out the dilemmas of their solitude in lonely and difficult places. Though these travelers almost always have companions on their journeys, there is only one voice that we really hear, and only one character that counts: the travelers themselves, characters so powerful, so twisted, so packed with sensibility that even the very striking places they travel through frequently have trouble competing for our attention.

Eric Newby is not quite so obsessed, and more important, he has a wife, the devoted but by no means docile Wanda, the woman who saved him from the Nazis in the Apennines during World War II. Wanda sounds like a great wife, but she is certainly, so far as that book goes, a great character, and it is the fact that there are two characters interacting with each other and with the cultures they encounter that causes *Slowly down the Ganges* to seem as much roman-fleuve as travel narrative, particularly since there is the *fleuve* itself—a great river holy to millions—to make even a third character. The book often makes the point that I just made about American roads: that the river and the country it runs through constitute two different realities. The Newbys soon discover that villagers living only a mile or two from the Ganges know almost nothing about it, while the river men are similarly ignorant of conditions even a little distance up the shore. River and village, roadway and forest are two realities that seldom merge, however close they may lie to each other geographically.

Slowly down the Ganges, insofar as it is both the story of a journey and—at least in part—the story of a marriage, has textures that most travel books lack.

I would like some of those textures for this book, but it is doubtful that I can achieve them, because I have no Wanda. I know many ladies, some of whom might like a little trip now and then, but I know no one who would be likely to enjoy sitting in a car with me while I plunge eight hundred miles down a highway in a single day, not equipped with a Zagat and *not even stopping for museums. Particularly* not stopping for museums, the acquisition of a broadened cultural awareness not being the point of these trips at all. (Just today, counting only Kansas and Oklahoma, I zipped past quite a few museums: the School Teacher Hall of Fame (Emporia), the National Four-String Banjo Hall of Fame (Guthrie), the Southern Plains Indian Museum (Anadarko), the Museum of the Great Plains (Lawton), the Cowboy Hall of Fame (Oklahoma City), and of course, more serious establishments in Minneapolis and Kansas City.

The inconvenient—even distressing—lack of a Wanda means that I'm apt to be writing a one-character book, the one character being someone I have only modest and flickering interest in: myself. In travel books such as Robert Byron's *The Road to Oxiana* or Bruce Chatwin's *In Patagonia* there is a kind of dance of sensibility going on, one energetic and vigorous enough to hold our interest. Robert Byron's interest in Persian architecture is so keen that it engages us, though perhaps not so strongly that we feel obliged to rush off to Iran and inspect his favorite building, the great tower called the Gombad-I-Kabus, whose brickwork Byron found so wholly admirable.

Similarly, even though we may be less interested than Bruce Chatwin in the last days of Butch Cassidy and the Sundance Kid, his investigation of their forlorn end—if it *was* their end—gives his story a kind of a ringing motif as he makes his way through Patagonia.

The challenge of the solitary traveler is always the same: to find something *out there* that the reader will enjoy knowing about, or at least, that the reader can be persuaded to read about. Usually, if there is no one but themselves in the narrative, the great travel writers rely on the extremes to which the environment forces them to produce the interest: Antarctica, and the failure of Scott to beat it, in Apsley Cherry-Garrard's *The Worst Journey in the World,* or Arabia's Empty Quarter and the ability of the Bedouin to *just* beat it, in Wilfred Thesiger's *Arabian Sands.*

I don't think I'm likely to encounter anything so extreme as the snows of Antarctica or the dunes of Arabia along the American interstates. At least I hope not. But I want to drive them anyway, even Wandaless, just to see what I see. I merely want to write about the roads as I find them, starting in January of 1999 in Duluth, Minnesota, at the north end of the long and lonesome 35.

JANUARY

The 35 from Duluth to Oklahoma City

I ARRIVED IN DULUTH on a wintry Sunday night. When I came to the head of the 35, just north of town, Lake Superior was invisible under a skein of snow. Though I was at the headwaters of the 35 I was by no means at the top of the country: Thunder Bay, to the northeast, and International Falls, to the northwest, were each almost two hundred miles away. I drove a little distance up the north shore of the lake, but it was snowing so hard that I couldn't see the water. Returning to Duluth, I noticed a nice design of a Viking ship on the highway overpasses. Some believe the Vikings got into the Great Lakes and onto the western prairies. I'm skeptical about this, but a certain Viking vigor is everywhere evident in Minnesota—how about Jesse Ventura, their wrestler-governor? When I checked into my hotel I looked out my window and saw three Viking-sized youths tussling happily with one another as they waited for their bus. Though the temperature was about ten degrees they were in shirtsleeves.

My hotel had a rooftop restaurant which, if not gastronomically promising, was at least very convenient. I had polished off quite a tasty bowl of soup and was about to attack my salad when I noticed that a group of photographs (boats on a lake) which had been at my elbow was no longer anywhere near my elbow. They were advancing steadily on the next table, and so, it appeared, was the wall they were attached to. When I began my meal I had been looking out long windows at the high bluff above the hotel, where most of residential Duluth lives, but now, instead, I was facing the dark lake, in which a large freighter was anchored—or at least parked—not more than five blocks from where I sat. My first thought was that I must be drunk. Was I so unable to tolerate a high latitude that three sips of bourbon had reduced me to a state of intoxication? It didn't seem possible, and in fact wasn't the problem. The plain truth was that the restaurant was revolving. My friend Commissioner Calvin A. "Bud" Trillin, in his magisterial work *The Tummy Trilogy*, has a stern warning about the inadvisability of eating in restaurants that move—an admonition I had obviously, if inadvertently, disregarded. The restaurant was definitely moving; soon I was looking at the high bluff again. But I had already ordered my walleyed pike, which, when it came, was quite good, forcing me to reflect that Commissioner Trillin may never have squarely faced the problem of dining in Duluth on a Sunday night.

I had just been reading Simone de Beauvoir's letters to Nelson

Algren, strange chronicle of a mostly unlucky romance that had occurred in the great city of Chicago, at the bottom of the neighboring lake. Eating in a revolving restaurant, contra Trillin, reminded me of Algren's famous adage: "Never eat at a place called Mom's, never play cards with a man named Doc, and never sleep with a woman whose problems are worse than your own." From the evidence of these letters it's not clear that Simone de Beauvoir's problems *were* worse than Nelson Algren's. At least she was the more tolerant of the two.

Nelson Algren was once arrested for stealing a typewriter from the journalism lab of Sul Ross College, in Alpine, Texas. Rather than serve time, he was allowed to leave the state, which he had visited in the depths of the Depression, hoping to secure employment. He did secure employment of a sort, though not particularly gainful employment: he found himself shelling peas in an abandoned gas station in the lower Rio Grande valley; the Texas stories which are the product of this experience are crude but not lifeless. It's my opinion that the typing lab at Sul Ross ought to put a plaque on the wall as a tribute to Nelson Algren's Jean Valjean–like act. A valuable if flawed midwestern writer got his start there.

I read Algren avidly at one point in my life. His very prose seems to bespeak a gloom, a sadness, a deep midwestern frustration; his awareness, keen at moments, fails him often. He could only write of failure, never of success, and now seems a lesser Dreiser, with more wit but less depth than his master.

One of the most delicious moments in Norman Mailer's *Advertisements for Myself* is a brief account of being on an early morning talk show with Algren in Baltimore, in which Algren, when asked to name America's finest writer, named William Styron, with whom, at the time, Mailer felt a particularly sharp rivalry. Once they were back on the street Mailer asked Algren why he had said it and Algren allowed as how he had seen a picture of Styron and his wife, Rose, departing for Europe on the *Queen Mary*—the image, perhaps reminiscent of Scott Fitzgerald and Zelda, seemed to Nelson Algren to symbolize how an American writer should look. Mailer's wry comment was, "Two middle-weight writers had fought to a draw in Baltimore."

Algren's own Sartre-haunted visits to Europe probably never matched that single image of William Styron getting on a boat with his wife, Rose. What he had accidently revealed, on a talk show with Mailer, was that in his own eyes, he was still a hick shelling peas in a gas station in south Texas, a man who had to steal a typewriter from a college in order to obtain a tool with which to begin his life work, a work in which there is little that is permanent, though he does now and then catch clearly the tone of human sadness as it manifests itself in the gritty cities of the midwest, the very country that lay below me when I set out the next morning, at 6 A.M. on a Monday, with the temperature only a notch or two above zero.

It had stopped snowing during the night—the roadway was clear. In a few minutes I passed the brilliantly lit Black Bear Casino, just south of Duluth, where, gambling-wise, the week was getting off to a slow start. There were only about twenty cars in the vast parking lot. As I proceeded down the 35 I passed, in one day, billboards for about a dozen casinos, many of them situated just to the east of me, in wicked Wisconsin. That afternoon I noticed a big one on the Missouri River, as I was passing through

Kansas City. Judging from the Black Bear's empty parking lot, the midwestern casinos have not yet managed to banish the concept of time as effectively as the clockless city of Las Vegas. Most midwesterners, though willing and even eager to gamble away their money—something fun to do at last—probably still feel that on Monday morning one should be going to work, not to the craps tables or the slots. Of course, that may change. Casino gambling in the midwest is a very recent thing. In time the stern midwestern work ethic may break down, or at least bend. At worst the casinos provide midwesterners with a way to be a little wicked without traveling all the way to Vegas. They can be wicked close to home and still go to work on Monday morning, securing at least a little of the best of both worlds. Also, the casinos provide at least a modest amount of glamour in a notably glamourless place; perhaps community boredom will be a little less widespread, a little less likely to drip its poisons into family life.

As I sped south a skim-milk light began to spread itself over the forests and the fields. I was in the land of lakes—*mille lacs,* to use the term the *voyageurs* bestowed on it. On some of the *mille lacs* the ice fishermen were already out. Some of them were inside little frame huts but most didn't bother with the huts, since the temperature, by their standards, was mild—though I did see one fisherman, evidently cold natured, fishing out of the cab of his pickup, dangling a line into his hole. A couple of the fishermen were drinking coffee—I could see steam rising out of thermoses.

The rush hour had arrived by the time I reached the Twin Cities; the 35 bifurcates, requiring one to choose between twins. I chose St. Paul—older, seedier, a little blowsy, less conspicuously virtuous, birthplace of Scott Fitzgerald and workplace of Governor Jesse Ventura. In fact I passed right by the state capitol at about the time the governor would have been showing up for work. The traffic was so light that I passed through St. Paul without ever having to touch my brake. I crossed the Mississippi on an old narrow bridge; the dark water steamed at that hour like the thermoses of the ice fishermen, only more so.

I have spent a good bit of time in the Twin Cities as a book scout; as in most northern cities with long winters, there are plenty of books, mostly of a scholarly nature. I had considered visiting Melvin McCosh, a bookseller who had filled a vast old nursing home near Minneapolis with books, but to do so would have involved me with serious Minneapolis traffic, so I decided to pass on and catch Melvin McCosh another time.

St. Paul is more or less the northernmost point at which the Mississippi is easily navigable. The French priests, often the first Europeans to penetrate this region, were there by the seventeenth century, but it was in the middle of the nineteenth that St. Paul became one of the principal funnels of immigration onto the vast farmlands to the west. As a city it still has the virtues of the melting pot: easygoing, tolerant, with plenty of watering holes and lots of people watering at them; Minneapolis, by contrast, is intimidatingly spick-and-span.

One thing I noticed on the drive down from Duluth is that the adopt-a-highway program, perhaps a grandchild of Lady Bird Johnson's determination to have a beautiful, litter-free America, is nowhere more successful than in Minnesota. Virtually every one of the 145 miles between Duluth and the Twin Cities has been adopted by someone—either a civic group or just a family who had decided to keep two miles of roadway litter free in memory of a loved one. The same program exists in Texas, and in many other states, but in Texas it doesn't work. All the highways leading into my hometown—Archer City—have been adopted by some group or other, yet all of them are liberally sprinkled with litter. I have yet to see a single person picking up litter on the roadways leading into Archer City.

Minnesota, by contrast, is almost blindingly litter free. But—I couldn't help thinking—what about the can scavengers? How do they survive in Minnesota? The only reason the various entrances to Archer City are not entirely blocked by mounds of beer cans is that one old can scavenger is out there every day, harvesting the can in one-hundred-can sacks. What would he be doing if he lived in Minnesota? Waiting for the end on a street corner?

My casual intention, in thinking about these journeys, was to have a look at the literature that had come out of the states I passed through. For Minnesota there is not a whole lot. Scott Fitzgerald, though a native son, spent most of his life east of Princeton or west of Pasadena. His work seems to me to owe little or nothing to the midwest. Louise Erdrich lives in Minneapolis now, but most of her work is set well to the west, near the Turtle Mountain Reservation in North Dakota.

I knew, a little, Frederick Manfred, the very tall Frisian-Sioux writer (pen name Feike Feikema), the Minnesota *romancier* whose most enduring romance was probably with himself but who, nonetheless, among his forty or more books, wrote two or three that are pretty good: *Lord Grizzly,* about the grizzly-mauled mountain man Hugh Glass, and *Riders of Judgment,* about the Johnson County cattle war in Wyoming, being my own favorites. Most of his fiction is set in a region he called Siouxland, although tribes other than the Sioux lived in some of that country. The last time I saw Frederick Manfred, when he was in his early eighties, he was hale and hearty and his eye for the ladies was in no way dimmed—though it was to dim shortly thereafter, as a result of brain cancer, which killed him.

I had with me on this drive one of the first travel books to treat of Minnesota, George W. Featherstonhaugh's *A Canoe Voyage down the Minney-Sotar*, not published until 1847 but describing a trip Featherstonhaugh took in 1835 up the St. Peter's River, where he hoped to find copper deposits. His publishers insisted on calling the river the Minney-Sotar, in hopes of capturing the audience that Longfellow captured eight years later with *The Song of Hiawatha*.

Featherstonhaugh was a cool, frequently condescending English geologist who, like Dickens and Mrs. Trollope, found much to disapprove of in American manners and customs. Though snobbish, he was not dumb, and he wrote vividly of what he saw. Even in 1835 it was possible for a casual visitor to see that the fur trade had already thinned out the once vast abundance of game. Featherstonhaugh saw too that a dependence on the white man's trade goods would soon degrade and destroy the once self-sufficient Native American cultures. The *Canoe Voyage* contains the best descriptions we have of the wild rice marshes and the tallgrass prairies long before these prairies were broken by the plows of the many emigrants who poured through St. Paul in the middle of the nineteenth century.

I don't generally like northern places: too much gloom, bred of short days, cloudy skies, long nights, cold weather. But I do like Minnesota. There is an appealing energy there, and a corresponding sense of health. Minnesotans, insofar as one can generalize, evidently thrive on cold, like those ice fishermen who were out with the dawn. I would be the last person on the planet who would voluntarily rise before dawn to go stand on a frozen lake and fish through a hole in the ice, but I can admire the hardiness of people who do.

It may be that in this high latitude, if we seek joie de vivre in art, we had better look to movies and TV, in which case, for Minnesota, the definitive texts would probably be *The Mary Tyler Moore Show* and the movie *Fargo*. Though I'm a great admirer of *The Mary Tyler Moore Show*, I always thought it contained an unacknowledged geographic dissonance. Mary walks to work and throws her cap up in Minneapolis, but her colleagues in the newsroom, in their vivid homeliness, seem as if they really belong in St. Paul. (It's interesting that two of the dominant sitcoms of the late seventies, *The Mary Tyler Moore Show* and *The Bob Newhart Show*, both have title sequences that show good solid midwesterners walking to work, Mary in Minneapolis and Bob in Chicago.)

The Mary Tyler Moore Show ran for 147 episodes; for students of American manners it can take a lot of study; likewise *The Bob Newhart Show*, which was set in a larger city and thus offered us a more sophisticated mix of urban oddballs: Jerry, Carol, Howard, and the rest, all somehow held together by the vibrant normalcy of Emily (Suzanne Pleshette) and, of course, Bob Newhart himself, the shrink who somehow never gets it really right or totally wrong. Watch either show over its full length and both will appear to be about fairly solid personalities in whom, sooner or later, cracks appear. Lou, Mary, Bob, Murray, Georgette, Carol are all people who get the job done, despite the obstacles thrown up by Ted, Phyllis, Jerry, Howard, Mr. Carlin, and a few erratic visitors.

A more recent artwork bearing on the character of the Twin Cities is the Coen brothers' *Fargo,* a film which draws its title from the bleak little town on the Dakota-Minnesota border. At one point in her investigation of what is clearly some kind of a used-car scam, Marge, the small-town policewoman played brilliantly by Frances McDormand, has to go to the Twin Cities, where she allows herself a sophisticated night out with an old admirer, a divorced Asian businessman. In hope of seduction he takes her to a restaurant in the Radisson, as high as he can aim on the sophistication scale. Marge's night out, which leaves her with a not particularly good aftertaste, poses a question of some centrality for midwestern life. If you live in the midwest and crave a little glamour, where do you go and what do you do? How glamorous is the glamour you are likely to find, if you find any at all? What if the top, glamour-wise, is only a workaday hotel restaurant with a good smorgasbord? What does this say about what the *rest* of your life is going to be? Does this mean that midwesterners can't aspire to real elegance, can't dream of it, perhaps can't even conceive of it as something that might happen to themselves? Once, certainly, it was otherwise, at least to newly arrived emigrants. In the Norwegian O. E. Rölvaag's sorrowful novel *The Boat of Longing* (1933), Minneapolis is portrayed as a place of great gaiety and charm. How good Minneapolis looks may depend on how far out on the prairies you're coming to it from.

I was through St. Paul by 8:30 A.M. and crossed into Iowa about 10. South of the Twin Cities the farms began, many of them showing evidence of prosperity. The farmhouses were large, the barns immense, meant to hold lots of hay and provide shelter for the livestock when necessary. The farmhouses were mostly two-story affairs, with broad porches; they were meant to house people of some size and heft, many of whom were in evidence, starting up tractors and turning the dairy cattle out of their milking sheds.

The valley of the Minnesota River lay not far to the west—it was there, in 1862, that the bloody uprising of the Santee Sioux occurred. The Sioux were hungry; they appealed to the Indian agent, Thomas Galbraith, for food, which, as they knew, was there in the agency's storerooms. But Galbraith, fearful of losing his kickback if he released the stores before he himself had been paid, refused to distribute the food. His buddy Andrew Myrick, in a famous mot, remarked that if the Indians were hungry they could eat grass, or their own shit. Shortly thereafter the Sioux, led by Little Crow, attacked the Lower Sioux Agency. Andrew Myrick was among those killed—his mouth had been stuffed with grass. The Sioux were soon subdued by the U.S. Army, but not before several hundred settlers were killed. The terror that raged briefly on these prairies drove the line of settlement back one hundred miles. The victorious whites planned a great hanging, with as many as three hundred Sioux to be hung—but that ambitious number was eventually reduced to thirty-eight, the person who reduced it being President Abraham Lincoln, who despite the fact that he had a civil war on his hands, patiently reviewed files and condemned only those Sioux he was convinced had been guilty of rape or murder. After the eruption of the Sioux, settlement slowed a little, but not for long.

Just before leaving Minnesota I stopped for gas at an Amoco truck stop. While I was waiting to pay for my soft drink, in a line of annoyed truckers—annoyed because there was only one cashier and he had turned his back on them and was talking on the phone—I glanced at the nearby video rack and was surprised to see a number of sexploitation films from the early seventies, several of them featuring a young Pam Grier, as well as Sid Haig, the latter an actor even Quentin Tarantino has not got around to resurrecting. The titles immediately evident—I was as impatient as the truckers and didn't want to get out of line—were mainly women-in-Philippine-prison flicks, on the order of *The Big Doll House, Women in Cages*, and the like. Pam Grier is usually either an evil lesbian warden or a buxom victim; if she's the buxom victim, Barbara Steele may be the evil lesbian warden. To my amazement I noticed an early Pam Grier vehicle called *The Arena* (1973), which had been written by two Rice classmates of mine, the Louisiana poet John William Corrington and his wife, Joyce. In *The Arena* Pam plays a Nubian captive forced to fight as a gladiator with other captive women; reluctantly she comes to accept the concept of sisterhood. Fate has been kinder, on the whole, to Pam Grier than to Bill Corrington; she recently got to star in *Jackie Brown,* but he is dead.

As a trader in rare books and a longtime student of the swap meet scene, I'm always interested in specialized markets. Those seventies women-in-cages movies are not the type likely to show up at your neighborhood Blockbuster, even though some, like *The Arena*, are now packaged as Roger Corman Classics. But here they were in a truck stop near Albert Lea, Minnesota, for the entertainment of white truckers, a little example of just how weird our culture is.

The truckers, as always, were gloomy—it's hard to find a cheerful trucker. They may be the last free men left, the true cowboys of the road, but they are never free of the pressure to make their miles, and it's evident from their faces what a heavy pressure that must be.

Soon after being startled by the video rack I crossed out of Minnesota, having taken just four hours to descend through it from Duluth. Just before entering Iowa I crossed I-90, which goes east to La Crosse, Wisconsin, or west to Sioux Falls, South Dakota; it's a road I will meet again. Indeed, in my first day of driving I crossed three of the great east-west interstates, the 90 at the bottom of Minnesota, the 80 at Des Moines, Iowa, and the 70 at Kansas City, Missouri.

The fields I was driving between were snow covered all the way from Duluth well into Missouri. The fact most immediately noticeable as one moves from Minnesota to Iowa is that the latter has fewer trees and more wind. In Minnesota the snow lay evenly on the fields, but in Iowa the winds had created great dunes of snow, off which the bright sunlight glinted. The sense of northern abundance that the big farmhouses of Minnesota convey was less evident in this part of Iowa, though it can be found a little farther to the west. In the middle of this century, at the time of Nikita Khrushchev's visit, the farms of the American midwest were the most productive the world had ever known; but the economics of family farming—never easy—became particularly brutal in the seventies and eighties, when the cost of credit, upon which almost all American farmers depend, began to rise to heights that only the most solidly established farmers could afford. With land taxes, the cost of money, and the cost of farm machinery all rocketing upward, many farmers in the American midwest suddenly found themselves faced with bankruptcy, which is to say, extinction. In these decades the farm auction—those heartbreaking affairs in which a way of life that may have lasted three or four generations abruptly ends—became common throughout the midwest. Families who knew no life but the land suddenly found they couldn't afford to stay on the land.

These sad evictions are still occurring, and the changes they have wrought make it hard to find much cheer in some parts of Iowa. The family farms of the midwest, like the family ranches of the range west, were always the work of several generations. All through the twentieth century young farmers, like young ranchers, have grown up on home places that were very likely their grandparents'; they grew up assuming that the home place would always be theirs and that they—if they wanted—could continue doing the work their forebears did. But their forebears had two huge advantages over them: cheap land and cheaper credit. The recent generations of midwestern farmers have to cope not only with the slow failure of their farms but with a sense that they have failed their ancestors as well. Little wonder that their confusion and bitterness sometimes turn to suicidal despair.

Not much of this showed along the 35. For a time the differences between Minnesota and Iowa were fairly subtle. The farmhouses in Iowa were narrower, a little more austere. The skies were widening and there were no lakes or clumps of forest to provide a contrast to the bright plains. When, occasionally, I did see a small lake, there were no ice fishermen on it; in Iowa the pleasure was more likely to be hunting: pheasant, duck, geese, deer.

When I slipped briefly onto I-80 just north of Des Moines, I was moving at a speed that more or less matched the number of the highway, but I was immediately passed by three cars traveling so fast that it was almost as if I was stopped. One of them had a Wyoming license plate and the other two were bound for the Golden State itself. Their drivers, in the moment I had to glance at them, all looked grim. Very likely they had already seen more than enough of the American midwest and just wanted to be out of it by the end of the day, if not sooner.

In Iowa very few of the miles along the interstate had been adopted for the purpose of litter removal, and this dropped to none when I entered Missouri. I was running out of the snow— patches of dingy-looking brown prairie began to appear. I noticed a steady narrowing of food options at the exits from the interstate. In Minnesota, as I approached each exit, there would be three cheerful signs, telling me what gas, what food, and what lodgings I could enjoy if I pulled off. Minnesota provided an abundance of eateries, Iowa mainly just Burger King, and Missouri nothing: the food signs were often simply blank. The state had evidently put up the signs in a spirit of optimism not shared by the major fast food franchises.

As I drifted on down toward Kansas, stopping in Missouri only for a bowl of soup that had a strange, rubbery taste, the country looked more and more boondocky. The evidence of neatness and order that had been constant through Iowa and Minnesota gave way to slovenliness and disorder; the country looked scruffier and the people at the gas stations and Quik Stops more hard-bitten, meaner. But then, the midwest symbolized by amber waves of grain and large solid families has rarely been the midwest I've found. In the famous Grant Wood painting *American Gothic* it's the pitchfork I notice first. Is the farmer going to stick that pitchfork into a bale of hay, or is he going to stick it into his wife?

The midwest, one might note, has been the home, or at least the venting ground, for quite a few of our natural-born killers, including the original spree killer himself, Charles Starkweather of Nebraska, who, with his girlfriend, Caril Ann Fugate, startled the nation by inventing spree killing in 1958. Before they were captured in Lincoln, they killed eleven people, including Caril Ann's mother. The best movie to deal with this spree is Terrence Malick's *Badlands*. The killings that resulted in *In Cold Blood* took place in Kansas, which is also where Timothy McVeigh made the bomb that took out 168 people and ruined at least a thousand lives. The term "going postal," which means to blow one's stack and murder as many people as possible, we owe to a postal worker in Oklahoma, who went postal and killed fourteen people. The Posse Comitatus made its home in Kansas until the Aryan Nation types discovered that Idaho was even more congenial. An aggrieved loner in Killeen, Texas, went postal *and* military, shooting more than twenty people in a cafeteria.

People go postal everywhere now, even in a peaceful Scottish village, but the midwest still seems to produce more than its share of these killers. It may be that the midwest produces a distinct kind of disappointment, which, in some, becomes murderous resentment. I think this disappointment has to do with glamour— or rather, with the lack of it. After all, if the American media promises us anything, it's glamour. Every day, everywhere we look (except, alas, at home) we see glamour: on television, in the movies, in magazines. Glamour has become part of the American promise, but it's hard to scare up much of it in the midwest. Kids and adults have only to turn on the TV or pick up a magazine to see people in other places wearing better clothes than they wear, eating more interesting food than they eat, and having a better time than they're having. Kids in the midwest only get to see even modest levels of glamour if they happen to be on school trips to one or another of the midwestern cities: K.C., Omaha, St. Louis, the Twin Cities. In some, clearly, this lack of glamour festers. Charles Starkweather, in speaking about his motive for killing all those people, had this to say: "I never ate in a high-class restaurant, I never seen the New York Yankees play, I never been to Los Angeles. . . ."

There, in a sense, you have it. A longing for prettier things, prettier places, prettier people—for girls who look like the girls in movies or on TV—can drive some to murder. The glamour of Spago, or of Rodeo Drive or Madison Avenue, is as remote from the lives of most midwesterners as Camelot or Troy. What do boys do—the family farm lost—if they happen to grow up in little towns along the 35? Make a life of fixing cars? Unless they have enough scholastic or athletic ability to get scholarships and get out, that is the reality for many.

For people mired in the depths of what one might call the inner midwest—the very region I'm driving through—even the arrival of the outlet malls is a blessing. They may have to lead pretty dull lives, but at least now they can *shop*, see what's new at Liz Claiborne, or get their kids nice overalls at Oshkosh B'gosh. They can own at least a few of the things that people have in places where life looks more exciting. There may as yet be no Armani, no Comme des Garçons, not even a Gap in these little strips of stores by the road, but what's there is still better than what was there before. Midwesterners, after all, are mature compromisers when it comes to glamour or, for that matter, to life itself; if life means working in a feed store, fishing a lot, and occasionally getting drunk for fun, then so be it. But when the compromise sours—as with a Starkweather or a McVeigh—then the results are likely to be fatal for quite a few people.

Western Missouri, dun colored under a sky that had suddenly clouded over, induces a certain feeling of dullness. There is little to see, except a landscape that is neither quite woods nor quite plain. The night before, I had read most of a book called *The Heartland*, by the good but now almost forgotten writer Walter Havighurst, which argues that the classical midwest (the heart of the heart of the country, in William H. Gass's phrase) consists of Ohio, Indiana, Illinois. For Walter Havighurst, the west side of the big river was somewhere else, but that's where I've been all day, and it certainly seems midwestern to me.

The first real traffic I encounter is in Kansas City, which I passed through well before the rush hour. Even so, the mercilessness of Kansas City traffic was a bit of a shock. Kansas drivers are intolerant, not willing to make polite allowances for country travelers who may have drifted into the wrong lane; the merciless Kansans will merely push the inexpert back across—if not into—the Missouri River and make them try again. I got through without suffering that fate, but only because I had been there before and knew which lane to be in at the mergings of the 35 and the 29 with the mighty 70.

Kansas City itself—or perhaps I should say *its selves*—is an interesting city with which I've never quite clicked. What it certainly has going for it, though, is the Missouri River—without that great river the place would be as charmless as Wichita or Topeka. The riverfront, with its new casino, has the seedy grace of some of the other old river towns: Louisville, Memphis, St. Louis, St. Paul.

It could be argued that the Missouri River was the greatest interstate road of the nineteenth century, its only rival being the Mississippi itself. To appreciate Kansas City properly it would be necessary to hang around the waterfront for a while. Native son Calvin Trillin has been so insistent upon lauding the city's many splendid eateries that he has rather neglected its river-town ambiance. The Kansas City book that I like best is Edward Dahlberg's *Bottom Dogs;* Dahlberg's mother was a barber, with a barbershop under a bridge, at a time when barbering was not a common occupation for women. D. H. Lawrence praised the book, which was published just before he died. It was Dahlberg's first book; his style, which became, if anything, too grandiloquent, was at this point blunt and a little crude—but his material was vivid. *Bottom Dogs* reminds me a bit of Nelson Algren's first book, *Somebody in Boots*—they are two midwestern books in which a sense of life survives some deficiencies of manner. The vitality of the cities being described thrusts itself onto the young talent, the result being imperfect but powerful books. Otherwise, Kansas City seems mostly to have produced jazz musicians, barbecue cooks, and journalists.

Thanks to the insistent traffic, I'm happy when I reach the southern suburbs and can see the land opening again before me. The 35 north of Kansas City had been almost empty; though I was seldom the only car on the road, I was never in anything that felt like traffic, but once south of Kansas City, I was immediately reminded of why the interstates really exist: for the trucks. Interstate trucking at the present level of density is unthinkable on the two-lane roads. South of Kansas City I entered the rushing stream of NAFTA traffic—trucks in their thousands were spilling off the 70, bound on the long run to the border at Laredo.

The heavy flow of truck traffic toward Mexico means that the 35 south of Kansas City is not a road one can drive with much pleasure. For the next 175 miles, until I pulled off in Wichita, I was never out of sight of trucks, and usually one or more were both just ahead of me and just behind me. I was, by this time, wearying a little. I had been on the road nine hours, coming down from Duluth, and had told myself before starting that ten hours a day would really be driving enough. Ten hours on the interstates will take one at least six hundred miles, nothing like the heroic drives Neal Cassady and Jack Kerouac used to make, but still, far enough for most of us. I wasn't attempting heroics—I was just looking, after all, and I wasn't sure I wanted to fight the truck traffic all the way to Wichita. The problem was that America doesn't really arrange itself neatly into ten-hour drives. There *were* places to stop between Kansas City and Wichita, but the sun was bright again and I decided I wanted to see that nice seventy-mile stretch of prairie south of Emporia as the sun sank over it—so I went on.

Passing through Emporia with about an hour of sunlight left I was surprised to see a neat roadside sign announcing that Emporia was the hometown of William Allen White, a prolific editor, novelist, and columnist who was once a force to be reckoned with, at least in the midwest. Yet his many books now go unread (because they're unreadable) and his name is likely to resonate with very few of the motorists who pass that sign. I have several of his books for sale in my bookshop but cannot remember having sold even a single copy of any of them—still, his hometown thought enough of him to put up a sign, causing me to resolve to at least look into one or two of his books when I got home, in the hope of discovering what all the fuss was once about.

South of Emporia the sun, whose orange tip I had first seen that morning over a frozen lake in Minnesota, was sinking over Kansas, but sinking slowly. For another hour I watched it light the brown rolling prairies north of Wichita, prairies which the great trail herds of the nineteenth century had crossed. A little earlier those plains would have been covered with buffalo, the great southern herd grazing in its millions, with a mix of nomadic native peoples—Comanche, southern Cheyenne, Kiowa, Pawnee—following it, living off it. Go back another ten or twelve thousand years and there would have been woolly mammoths lumbering around the parts of the west where the buffalo lumbered later.

To a plainsman born and raised, as I am, that lovely unbroken Kansas prairie tugs at the heart and the memory, because it is rare now to see that much grazing land not pierced by oil wells or torn by the plow. In my boyhood we once or twice summered cattle in that part of Kansas. I remember once slightly embarrassing my father when we were out horseback with some cowboys, looking over a pasture. The unfamiliar and, as it turned out, unpredictable pony I had been assigned, after plodding along sullenly for several miles, suddenly gave a great bound, jumping one of the many rivulets that water those prairies, dumping me squarely in the middle of the small stream. There wasn't much water in the stream, I didn't get very wet, but the sense of intense embarrassment lingered, leaving me with a slight distrust of everything Kansan except the fine prairie itself: the Osage prairie, as it was once called.

The plains sunset was well worth the extra hour of driving. A wheat-colored light seemed to fill the whole west, developing purplish tinges along the edges, like bruises on the sky, as night came on. Before sunset I was several stories up, in the Wichita Marriott, looking south to the line of trees that marked the course of the Arkansas River—the river on whose banks my character Wilbarger dies. Wilbarger was the well-educated Yankee who called Augustus's bluff about the Latin motto on the sign in *Lonesome Dove*. He is one of my favorites, in a book that has 115 named characters, not counting the horses. It is sometimes the minor, not the major, characters in a novel who hold the author's affection longest. It may be that one loses affection for the major characters because they suck off so much energy as one pushes them on through their lives. Wilbarger read Milton by the campfire and then went off and got himself killed in a shoot-out with the dreadful Suggs brothers, on the banks of the Arkansas River, whose steaming waters I crossed the next morning, when I left Wichita.

South of Wichita there was still some lovely prairie to see, but I couldn't see it because a thick ground mist hid the prairies for 150 miles and thickened into true fog as I temporarily took my leave of the 35 and slid off onto the westbound 44 on the northern outskirts of Oklahoma City.

Where Oklahoma City was concerned, I was glad of the fog, O.C. not being a place that offers much to the eye, or to the stomach either. It is a community born of and sustained by oil—at one time oil derricks crowded so close to the state capitol that an arcade of wooden awnings had to be constructed to keep the legislators from being splattered on their way to work. At one point in the city's history, in order to rake in as much oil revenue as possible, the civic authorities spread the city limits so generously that this relatively small city had the same surface area as Los Angeles, some 640 square miles.

Cloaked in fog, Oklahoma City looked better. I crossed the great river of interstate 40, carrying people east toward Little Rock or west toward Amarillo, Albuquerque, and California. The 120 miles of I-44 that I needed to travel, before the freeway peters out south of Lawton, are not very interesting, though I was cheered by the sight of the bumpy Wichita Mountains as I came into Lawton, an army town mostly noted for having been the last home of two nineteenth-century Native American leaders, Geronimo and Quanah Parker. Both were remarkable men, but Quanah managed the more remarkable transition. He was, in his youth, a formidable war chief, leading many raids; he was almost killed at the second battle of Adobe Walls—yet he survived, surrendered, and led his people, the Comanches, into a fairly stable relationship with the white government and the twentieth century.

Geronimo, by the time he was housed at Fort Sill, really had very few people—many of the eighteen warriors who had surrendered with him in 1886 had died in captivity. Geronimo survived twenty-three years as a prisoner, nineteen of them at Fort Sill. Though he longed for his native desert and petitioned every white leader he could find to send him back, he was never allowed to return to Arizona. Finally one day he got very drunk, spent a cold night outside, and died of pneumonia. In his last years he and Quanah had formed a friendship—two men who had seen their time, and their people's time, end.

Lawton is a tough town. I saw no reason to linger in it, but I did stop to buy a candy bar on Gore Boulevard, the street named for

Gore Vidal's grandfather, the blind senator. I crossed the sandy Red River near the town of Burkburnett, site of one of the century's more chaotic and colorful oil booms. The clock on the bank in Archer City said 12 noon exactly when I pulled up at the door of my bookshop. I had been on the road only a day and a half and would have welcomed another day, but there was no route that seemed particularly convenient. I had traveled exactly eleven hundred miles since leaving Duluth.

FEBRUARY

The 35 from Dallas–Fort Worth to Laredo

ELEVEN HUNDRED MILES, Duluth to Archer City, only made me restless to get back on the road. There was no need to hurry, yet I felt like hurrying. Fortunately I had to go to Austin that weekend, to visit my grandson; I also had to exchange rental cars at the Dallas–Fort Worth airport. Both Dallas and Austin lie along the 35. Since I was going to Austin anyway, I thought I might as well complete the road—excepting the somewhat dismal segment between Oklahoma City and Dallas.

The most interesting thing that ever happened to me in southern Oklahoma happened when I was a boy. My backwoods uncle Jeff Dobbs took me deep in the woods, to the cabin of an aged Choctaw preacher, an old man said to have the power to draw out tumors. In his small cabin there were two long rows of Mason jars, each containing a tumor that had been drawn out. It was dim in the cabin. I couldn't see what was in the jars very clearly, but it definitely wasn't string beans or pickled peaches. I was very impressed and not a little frightened. Uncle Jeff knew a few words of Choctaw—listening to him talk to the old man was when I first realized that there were languages other than English.

More than fifty years after I peered at them in the gloom of the old preacher's cabin, the shelves of tumors reappeared in *Pretty Boy Floyd,* the first of two novels I wrote with Diana Ossana. This time "the cancers," as they are referred to, appear as decoration in a backwoods honky-tonk.

When I left the Dallas–Fort Worth airport I slipped quickly through Arlington, not so much a city as an area of confusion that manages to combine the worst features of Dallas (just to the east) and Fort Worth (just to the west). My bitter dislike of Arlington goes back ten years, to a day when I embarrassed myself by getting hopelessly lost in it while attempting to take the world-famous, globally traveled author Jan Morris to lunch. Not long after this disgrace I complained about Arlington in a novel called *Some Can Whistle,* but nobody read that novel so no one heard my complaint. Passing through it again I noticed that it had become even uglier—my complaint, if anything, had been understated.

I cut through Arlington in order to get to I-20, the midsouth interstate that slides just south of Fort Worth and would reconnect me with the 35. The 20, whichever way you travel it, leads to nowhere interesting. The short stretch of it I drove this time was a road badly in need of adoption by some litter martyr—there was enough litter just along a fifteen-mile stretch to start a new landfill. Locals apparently didn't notice it, perhaps because they were in such a hurry to get to one of the huge, commandingly ugly malls just off the interstate. The most eye-catching billboards along this stretch were for microsurgical vasectomy reversal, a growth industry in Texas, evidently. A lot of good old boys must have come to feel that, after all, they want families rather than just fucking.

The 35 between Fort Worth and Austin is not a pleasant drive. It's an old, crumbling, badly banked freeway which carries much too much traffic, all of it going much too fast. Truckers bound for Mexico duel constantly with students in subcompacts, impatient to gain a car length or two; the kids dart from lane to lane like minnows. I slipped past Alvarado, hometown of Terry Southern, but by then I was in traffic so fierce that I had no time to think about the man or his books. A pack of coons once invaded my ranch house, which, at the time, held over ten thousand books, behind which my exterminator had planted big green chunks of rat poison. Though fatal to rats, this particular poison was evidently candy to coons, who knocked many books off the shelves in order to get at it. The only book really damaged by this invasion was my copy of *Red Dirt Marijuana,* which had been almost entirely devoured. I wrote Terry Southern about this incident and got an amusing note back.

Had I been in a museum mood I could have inspected two dandies in Waco—the Texas Ranger Hall of Fame and Museum or the even dandier Dr Pepper Museum. Waco, though the home of the Texas Baptists, seems to have been forced by the times and the mores into a slightly more tolerant attitude toward sin. A billboard for a local club offered total nudity, while several others promised vasectomy reversal. Waco is the town where the nineteenth-century gadfly William Cowper Brann, editor of a rag called the *Iconoclast,* was shot dead in the street in retaliation for his fierce attacks on Baptist behavior.

In Austin, while enjoying a lively debate with my grandson Curtis McMurtry—all debates with Curtis are lively—I read the *Times,* where I learned, to my amusement, that Norman Podhoretz had once said, "Fuck you," to Jacqueline Kennedy Onassis. This revelation came in a discussion of why he had ceased to be friends with Norman Mailer. I found this particularly interesting because it was Norman Mailer who had once turned to the then-queenly Diana Trilling and asked, "What about you, smart cunt?" Neither lady seems to have been particularly displeased by these vulgarities, though both in time got their own back with Mailer and Podhoretz. It makes one wonder why Mailer and Podhoretz aren't still friends, since both are good at employing less than respectful speech to get the attention of queens—a basic skill that will long outlast their political opinions.

My views on the respective flaws of Austin and San Antonio I set down much earlier, in a book called *In a Narrow Grave;* about those two cities I have nothing to add. What has changed is that the pleasant stretch of prairie that once lay between Austin and San Antonio has been filled in with San Marcos, New Braunfels, a large outlet mall, and a racetrack. This is what one might call urban scatter, since it lacks the full density of urban sprawl. The country is neither full nor empty. Five or ten miles of nice country will occur, but then there will be another little clump of businesses, squatting vulturishly beside the road, waiting to eat you.

South of San Antonio this stops and the 35 changes dramatically for the better. It ceases to be a bobsled run and becomes a broad, comfortable, well-tended highway, with a big median and ample shoulders. I am now in the narrowing southern cone of Texas. The brush thickens and the towns thin out—the brush is *so* thick that one can see that the pioneers might have felt it just wasn't worth it to hack out towns in such a place. Mesquite, chaparral, and prickly pear are so entwined that one can scarcely see into them. It is an environment in which rattlesnakes, armadillos, and javelinas abound. At one time the cattle that roamed in this region were the wildest and most dangerous animals to be found. The brush country is the opposite of a rain forest—it's a sort of dry forest.

I first went to Laredo, the border town, with my father when a boy of five, while he was on a trip to inspect a cattle herd at the behest of some bank. We had plenty of mesquite in our country, four hundred miles north, but nothing like what I saw that day as we drove into a kind of south Texas heart of darkness. I was astonished and a little scared: two steps into those thickets and one could disappear forever; later that day, in pursuing his duties, my father mounted a horse and did just that, leaving me to fret in the car. That night we stayed in a cow camp that had somehow been beaten out of the brush. About suppertime a vaquero rode in with a dead javelina across his horse's rump; our supper consisted of wild pig and frijoles. The vaqueros, to my young eyes, looked like wild men; apart from the frijoles they seemed to live on what they shot: pigs, deer, now and then a turkey. What I felt then about deep south Texas—that it was a wild place filled with desperate men—is pretty much what I still feel.

The next night my father and I stayed in a hotel in Laredo whose bedbugs were so fierce that we eventually gave up and slept in the car. At some point we crossed the bridge into Nuevo Laredo—

there is a photograph of me sitting on an old steer with lyrelike horns, perhaps a zebu. I have no memory of the event at all; only that photograph convinces me that it happened. It would have been the only time in his life that my father set foot on foreign soil.

In the seventies I went into the border country often. Danny Deck, the hero of *All My Friends Are Going to Be Strangers*, flees there the night his daughter is born; she reappears eighteen years later in *Some Can Whistle*. At the end of the first book Danny drowns his second novel in the Rio Grande, near the town of Roma, where some of *Viva Zapata* was filmed. I still feel drawn to the stretch of country between Laredo and the sea, but drawn to it ambivalently. I've never been comfortable there: it's a hard, scary land. We took it from Mexico by conquest but—culturally and morally—have never won it. The poor Hispanic people who live there were kept for decades under the heavy heel of the Texas Rangers—nowadays it's the heavy heel of the Border Patrol, which is like an occupying army, ever at war with the sometimes better equipped army of the drug dealers on both sides of the river. The Hispanic poor from deep in Mexico or even beyond Mexico— from Guatemala and El Salvador—are caught between these two hostile and militant forces, and yet, despite all risks and impediments, keep flowing north. High tension, brutality, and violence have been part of border life ever since the Rio Grande became a boundary between two nations, rather than just a river that people crossed and recrossed when they needed to, or when they just felt like it. The various police agencies that have tried to stop this natural, northbound flow of people toward what they hope will be a better life only degrade themselves in the process, since no human can legitimately be denied the desire to survive and help their families. The great border crossing points are heavy with the apparatus of international commerce, but for most of its length, both Texans and Mexicans find it hard to take the muddy little river seriously as a border. *Lonesome Dove* opens with a cross-border raid; the tensions expressed in it are not resolved. Most Mexicans still feel that they have an innate right to be on the north side of the river, where their grandparents or great-grandparents lived. The Border Patrol can deport them, but it can't extinguish this feeling.

It takes less than two hours to slide down the 35 from San Antonio to Laredo, the latter a town that has changed much since my father and I visited it in 1941. As one approaches it vast warehouses begin to appear, several of them as large as any buildings I have seen anywhere. These are the great distribution centers where goods going south or coming north are sorted and dispatched on the appropriate carriers. There are several places near the highway, ez-on, ez-off, where the truckers can get their rigs washed. It would appear that international freight forwarding is now Laredo's raison d'être. As I drove closer to the border itself I saw that every imaginable chain business had secured itself a prime spot near the roadway, in a density that had not occurred for the whole fifteen-hundred-mile length of the road. Here one can stuff oneself one last time with rapidly prepared American food, before plunging into the land of the Bad Amoebas. I began to fear that the old Laredo had disappeared completely, to be replaced by a kind of Chain City, but I went on anyway, to where the 35 ends, about a mile from the international bridge. A gas station was positioned at the south end of the road, just as one had been at the north end. It was not until I reached the very end of the great road that I found the old Laredo, the cluster of taco stands and open shops offering a vast assortment of tacky border town goods. Of course, all that cheap jewelry, those belts, those ponchos, those sombreros, those pots and rugs might look highly desirable to people from Duluth. Though the little border strip in Laredo is not radically different from what one would find in Del Rio, Juarez, Nogales, or Tijuana, here the palm trees give the scene a sort of dusty Middle Eastern tint. The street stops feeling like Texas and feels, for a moment, like Morocco. Suave hustlers with lots of gold chains beckon from every doorway. At the north end of the 35 there's a lake, and at the south end there's a souk. A lot of cheap pots and faux-Zapotecan rugs being offered will eventually adorn the modest bungalows of the snowbirds who come south every year to the warmth, bungalows that might be in Milwaukee or La Crosse.

It was a warm day, some eighty degrees hotter than it had been when I pulled onto the 35 in Duluth. I parked and strolled a few blocks through Laredo, a town where almost everyone is willing to bargain for almost everything: sex, drugs, palm readings, pots, silver chains. I had a couple of excellent bean burritos at a little stand: when in doubt along the border it is best to trust the bean.

On a bright sunny afternoon in Laredo the border offers no menace. But if I were to go up- or downriver, into the empty spaces between communities, it would be different. I am rarely uneasy anywhere on the American road, but I do feel some unease along certain southern stretches of the Rio Grande, where the drug dealers have begun to drive out the small ranchers. The lower Rio Grande valley is now too heavily settled to feel threatening, but upriver it's different. There are stretches where the country seems little more settled, and no more peaceable, than it was when McNelly's Rangers came along and attempted to tame the Nueces Strip—the lawless area between the Nueces River and the Rio Grande.

At the Border Patrol station, once I had headed back north on the 35, the drug-sniffing dog had his paws on my car before I had even stopped, but the patrolman immediately yanked him back and waved me on. The dope dogs are a sophistication Captain McNelly or his successors in the Texas Rangers didn't have—the early Rangers were too busy avoiding violent death, or dealing it out, to worry about such things as drugs.

Once through the Border Patrol station I was faced with a dilemma. I was through with I-35, but I was also four hundred miles from home. I might have gone east to highway 281, the road whose murmurs had lulled me in my childhood, but getting to the 281 and then traveling up it would have added two hundred miles to my trip. I chose instead to travel north on the underappreciated highway 83, a beautiful, straight, empty two-lane road that begins in Laredo and slices on north into Canada, bisecting the central plains. Along the whole length of the 83, from the Mexican border to Melita, Manitoba, the largest town a driver will encounter is Abilene, Texas, whose population is just over one hundred thousand.

It was a Sunday afternoon in winter. There was almost no traffic on the 83. The western horizon, distant to begin with, became more distant as I drove north. I sailed past vast plantations of prickly pear and mesquite. In a few places mesquite, chaparral, and prickly pear had been bulldozed into great piles; thickets were going to be turned into farms, though no farms were in evidence yet. For about one hundred miles the 83 runs parallel to the northern course of the Rio Grande; a little town such as Carrizo Springs seems like a Mexican village whose residents have had the good or ill fortune to end up on the north side of the river.

As I went north, though, onto the Edwards Plateau, the border ambiance gave way to that of the nouveau ranchers. For the next 150 miles I kept passing monumental ranch entrances whose archetype was the Arc de Triomphe, with welded sculptures of cutting horses, or perhaps longhorns, beneath the arch. It is not likely that the hard-bitten old German sheep farmers who settled this area would have approved of—much less paid for—French pillars or welded sculpture.

The highway was so empty that I was able to travel at quite illegal speeds—a good deal faster even than I had traveled on the broad 35. I had to slow down a little for a curvy fifty-mile stretch of road that cuts through the unfashionable western edge of the hill country—unfashionable because it's really too far from the Austin airport, and is thus inaccessible to the weekend rich.

I have always been a little puzzled by the popularity of the hill country. The soil is too stoney to farm or ranch, the hills are just sort of forested speed bumps, and the people, mostly of stern Teutonic stock, are suspicious, tightfisted, unfriendly, and mean. Even the foliage is mean, releasing a steady cloud of powerful allergens into the air. Perhaps the very fact that the country is too stony to farm successfully makes it even more perfectly suited to be an ideal of rusticity. Willie Nelson's famous song "Luckenbach, Texas" has probably lulled many people into thinking that Luckenbach and its neighboring communities are friendly places, when in fact they aren't. Some of the hill country rivers are nice—they make good places to establish summer camps for city kids; otherwise about the only species that flourishes in this limestoney environment is the goat, colorful varieties of which can be seen dotting the hills.

Just north of Ballinger I got a great lift: I came over a hill and the sky suddenly swelled to infinity—I could see at least fifty miles from horizon to horizon. This huge spread of sky was with me all the way to Abilene. The relationship of sky to land in the plains west is one I find continually affecting.

Though Abilene is only about two hours from my home I decided to spend the night there, mainly because it promised to be a clear morning and I wanted to see the Brazos valley at sunrise, rather than sunset.

This I did, coming up through Albany and Throckmorton. Between the two small towns, as I was almost to Fort Griffin, I passed a small metal sign propped against a post by a narrow dirt road. The sign was hand lettered on a piece of metal siding and merely said "Matthews Ranch."

After the score or more of ugly Arc de Triomphe ranch entrances to mostly faux ranches that I had passed on my way up from Laredo, it was refreshing to see that such a modest, even casual sign marked the entrance to one of the most famous ranches in the American west; that would be the Lambshead, whose patriarch, Watkins Reynolds "Watt" Matthews, had died the previous year at the age of ninety-eight. No showiness is sought or allowed on the Lambshead, one of the few true frontier ranches that is still held by the same family and still functions as a ranch. The Lambshead was established in the 1860s, when Indian attacks were common events.

Watt Matthews's formidable mother, Sallie Reynolds Matthews, wrote a famous ranch memoir called *Interwoven* (1936), in which she passes rather lightly over the dangers of the early years on the Clear Fork of the Brazos, though the dangers were real, frequent, and often desperate. The Comanche and the

Kiowa made the most of the Civil War years, when the white militias were, to say the least, distracted. Sallie Reynolds Matthews was mostly self-taught and was no doubt frustrated by the limited opportunities for learning that she faced in her early years; determined that at least one of her children would know something other than the cattle business, she insisted that her son Watt go to Princeton. Watt Matthews was known for his reluctance to leave the Lambshead, but he eventually became an enthusiastic Princetonian, attending all the reunions as long as he was able, and he was able for three quarters of a century. He would have been at Princeton just after the time of F. Scott Fitzgerald and Edmund Wilson. The rest of his life he devoted to the Lambshead; he never married or, so far as I know, even flirted. A ranching family is lucky to produce one member with his tenacity and level of focus. I don't know what will become of the ranch without him, but it would not surprise me to drive along that road in a year or so and see an Arc de Triomphe entrance to the famous Lambshead ranch. If so, it will be a change of which Watt Matthews and his strict forebears would have greatly disapproved.

Only a little farther, just before crossing the Brazos, I came to Fort Griffin—no doubt the Lambshead's proximity to that army post was helpful, perhaps even critical, in the scary early years. The Fort Griffin area was one of the last shipping points for buffalo hides; part of the great southern herd once grazed on what became the Lambshead. The writer Max Crawford has a fine novel, called *Lords of the Plain,* some of which is set near Fort Griffin, in a place called Hidetown. I moved the same place to New Mexico and called it Greasy Corners. In actuality Greasy Corners is in the Oklahoma panhandle. Between the arrival of the first whites into this country in the 1840s and the eventual defeat of the Comanches in the 1870s the whole area watered by the several forks of the Brazos—the Clear, the Salt, the Double Mountain, the Prairie Dog— was hotly disputed territory. Much of it lay directly on the Great Comanche War Trail, the raiding route that the Comanches had been using for a long time when they went to Mexico in search of booty or slaves. Captain Call and Captain McCrae, my Rangers in *Lonesome Dove*, would have been thoroughly familiar with it, as was Ranald Mackenzie, the young army officer who finally dealt the Comanches a decisive blow in September of 1874, after which he was sent north to help Miles and Crook fight the Sioux, the Cheyenne, and the Utes. Ranald Mackenzie even made two quick forays into Mexico, to subdue some hostile Lipan Apaches.

Mackenzie was a highly effective officer, one of the most skilled and determined to fight on the plains frontier. But he was not a happy man. Just before he was to marry, in 1883, he went crazy and spent the remaining six years of his life in an insane asylum in New York State. Ranald Mackenzie's insanity is one of the strange, haunting mysteries thrown up by the frontier conflicts. Many pioneer women went crazy, and it was not hard to see why; the women were not necessarily overdelicate, either. The living conditions were just too bleak, too isolating. But the insanity of Ranald Mackenzie, one of the most disciplined and successful officers to participate in the campaigns of the plains frontier, is evidence that the price of winning the west was not simple and not low, even for the winners, not when one considers that Ranald Mackenzie, the soldier who took the surrender of Quanah Parker and the Kwahadi Comanches, ended his days in a nuthouse, in 1889, not long before the massacre at Wounded Knee.

FEBRUARY

The 10 from Jacksonville to Lafayette.
The 49 to Shreveport

FLYING INTO THE Jacksonville airport is like landing at a minia-turized, brand-new, spick-and-span LAX. The sun is shining but it isn't hot; everyone is perky; and the arriviste can even catch a pleasing whiff of sea air. It's bright but coolish; north Florida attire runs loosely to shorts worn with windbreakers; the nice air-port is a petite, well-scrubbed version of the big one at the other end of the road.

In the course of my travels I've stayed, as I've already said, at more than two hundred Holiday Inns, from which the last thing I expect is a surprise: but I got one at the Holiday Inn near the air-port in Jacksonville, which had a sheriff's office smack in the middle of the lobby. This one was empty of sheriffs at the time: the only thing on the neat desk was a copy of the swimsuit issue of *Sports Illustrated*. But it was a sheriff's office, nonetheless. Lots of small conventions hit Jacksonville—a modest one was convening in the lobby even as I was checking in. Later in the evening I heard unmistakable signs of rowdiness but did not peek out to see if the sheriff had arrived to control it. As I was leaving the next morning I asked the young lady at the desk if it was necessary to pack the sheriff's office with revelers. "No, but they drag off plenty of drunks," she allowed.

THE WESTERN END of the 10, in Santa Monica, is quite dramatic. The big road that has just surged through downtown L.A. drops you right off the palisades, almost underneath the Santa Monica pier; you plunge downward onto the coast highway, bound north to mystic Malibu, with the great ocean bordering the road. Just a block or two north of where the 10 ends, strollers on Ocean Boulevard can examine a plaque to Will Rogers, the much-beloved radio philosophe. The plaque is at the western end of old route 66, the road Rogers took on his route out of Oklahoma and into the hearts of millions.

Nothing that dramatic awaits one at the eastern end of the 10, which begins rather indecisively (or ends, if you prefer to think west-east rather than east-west) just about a mile from downtown Jacksonville. I wanted it to begin at the Atlantic beaches and was disappointed when it didn't, although I was able to proceed to the Atlantic along the "little 10" (Atlantic Boulevard), a road through east Jacksonville that finally took me to the beach. I had wanted to see the sun come up over the Atlantic, but thanks to many stop-lights and much traffic had to content myself with watching it come up over the estuary of the St. Johns River, which was lovely enough. As a bonus I got to see a little of what remains of old, funky downtown Jacksonville.

One beneficent characteristic of oceans is that they tend to relax the people who live by them. Worldwide, in my experience, littorals are apt to have a wait-and-see, gentle seediness. Even the most manicured littorals—Malibu, Lanai, Cap d'Antibes—have a touch of it. The force of huge money may keep the beaches look-ing neat for a mile or two, but neatness and the seashore don't really go together; only one hundred feet or so past the great resorts the old casual seediness begins. The structures lean a lit-tle, and the people who inhabit them don't worry too much about dress codes. The sea, eternal, shelters and soothes them, saves them for a time from the ambitious strivings that drive people who live far removed from the ocean.

By the time I finally got to Buccaneer Beach the sun was up, the day cloudless. I breathed in a little Atlantic air and headed west. Just as I entered the big 10, I passed a huge flea market, which was cheering. The swap meet culture, of which I have long been a student, is now truly continental: there's a fine swap meet at the Rose Bowl, at the other end of the 10. It was in the seven-ties that swap meets really began to be popular in America; attics, closets, and garages throughout the country began to disgorge their contents onto tables in the parking lots of strip malls—usu-ally older, dying strip malls. Swap meets now are much more pro-fessionalized, but they still yield amazing harvests of goods. Love ends, life ends, but the flow of objects goes on forever.

I had gone only a few miles west on the 10 when I noticed something a little unusual: the rest stops I passed all had signs conspicuously promising security services after dark. Later in the day, after I crossed a little tooth of Alabama and entered Mississippi, the rest area signs offered twenty-four-hour security. These offers are no doubt a response to a series of robberies and murders that occurred at rest stops a few years back, though the offer of after-dark or all-day security is not one that holds for the other states the 10 goes through: Louisiana, Texas, New Mexico, Arizona, and California. In those places, if you have to pee or want to walk your pet, you're on your own. I long ago stopped using official rest areas—even if you escape muggers and murderers, you're sure to be assailed by road rats, hustlers, or religious cranks.

Despite the perils lurking at the rest stops, it was a fine morning for a drive—unobstructed sunlight lasted all day. I saw the sun come up over the St. Johns River and saw it set on the Red, in Shreveport, Louisiana. What I didn't get, and didn't expect, was much variety along the roadsides. I was looking at seven hundred miles of southern pine, with only now and then a patch of swamp to break the monotony.

I was in the mood for blank driving, accompanied by minimal thought, if any. I decided to forgo an examination of my rather complex and largely negative feelings about the south until I was coming down the 75 or the 55 or some other southern road. I crossed the south many times when I was young: by bus twice, and by car several more times, grinding through it on the old, cracked, two-lane roads that connected the wretched, racist small towns. Once in Georgia I was spoken to sharply by an elderly waitress who took the fact that I didn't want grits as evidence that

I was probably a northern agitator of some kind. She even mentioned this damaging fact to a deputy sheriff who happened in. I think she hoped my dislike of grits would be cause for arrest, but the deputy just grinned. "I guess it ain't a crime to hate grits," he said. "I hate scrambled eggs myself."

The Florida panhandle runs for almost four hundred miles, from the Atlantic coast to Alabama. Crossing it required a morning, though a pleasant morning. The pine forests looked their best in the bright sunlight. The 10 slices north of Tallahassee—it touches no cities until it meets Mobile. Escambia Bay, at Pensacola, is a reminder that the Gulf is near—I have been running only about fifty miles north of it for most of the morning.

Then I'm in Alabama for an hour—microsurgical vasectomy reversal seems to be big there too. To the north Alabama swells out, but in the bottom it narrows like a fang. What is clear as I approach Biloxi is that riverboat gambling has made it all the way down the center of the country, from northern Minnesota to the Gulf. Biloxi is big time, able to import entertainment on the order of Harry Belafonte, the Pointer Sisters, and Engelbert Humperdinck.

The country I drove through all morning, the Florida panhandle, relates more naturally to the states just above it—Georgia and Alabama—than it does to the semi- or subtropical peninsula that extends south to within one hundred miles of Cuba. North Florida is not the Florida of popular imagery: that would be Miami, West Palm Beach, Cape Canaveral, Orlando (which is to say Disney World), and Key West—that's the Florida of *Miami Vice*, South Beach, Versace, Tennessee Williams, Hemingway. Tallahassee, the state capital, has more in common with Valdosta, Georgia, or Montgomery, Alabama, than it does with Miami.

Orlando, thanks to Disney and the space center, is now the Capital of Clean, the place where families go, whereas Miami, which David Rieff has called the capital of Latin America, is the place where swingers go, where the music is loud, sexual peculiarities tolerated, and drugs plentiful.

I don't find present-day Miami very different from the Miami I knew in the early seventies. It's not so much different as just more so. The one time I stayed in the Fontainebleu an Irish waiter wheeled in my bacon and eggs one morning and, dead drunk, fell facedown in my eggs.

But for the moment, I'm in the panhandle, not the peninsula; then, briefly, I'm in Alabama and Mississippi and will soon be in Louisiana, where a choice must be made. Do I dip down through New Orleans, going down and back up with the 10, or should I proceed straight west on the 12, a nice cousin of the 10 that eliminates the tortured struggle across Lake Pontchartrain and into the New Orleans traffic?

I've been to New Orleans often and find I'm not in the mood for it this morning. Long ago, innocently, I went there hoping to ride the Streetcar Named Desire and wound up sleeping in my car for three nights—I had not realized it was Sugar Bowl week, which meant that the hotels insisted on a three-day minimum, which I didn't have. I wandered about the places where Faulkner, Sherwood Anderson, Tennessee Williams, and others had lived but kept straying out of these hallowed places into what seemed like only another poor southern city.

The most likely reason for crossing the lake would be to eat at one of the city's fine restaurants, but I have made no reservations and am not a foodie to that extent. I don't require Paul Prudhomme, and besides, I know that as soon as I get to the Bayou Teche there will be a little Cajun man parked beside the road in a pickup who will sell me excellent boudin sausage and perhaps a nice cup of gumbo (sure enough, he was there).

Long ago there had been a rather curious bookseller in Slidell, Louisiana, a man who specialized in very expensive erotica and equally expensive books about cats—the two tastes seem often to go together. I could not afford his books but it was fun to see his cats, haughty Abyssinians who slouched elegantly about the premises.

Louisiana traffic is not so much homicidal as suicidal. When I first drove to New Orleans, on the horrible two-lane highways of the day, I lived in terror of being casually brushed into a bayou by one of the many oil trucks hurtling through the swamps. This was not paranoia; it happened to many an unwary driver, though sometimes it was the oil truck that went into the swamp. The highways are wider now—at least the 12 was pleasantly wide— but people in Louisiana still drive as if they don't care whether they live or die. I rejoined the 10 at Baton Rouge, where I got to cross the Mississippi at my favorite crossing spot. It was a far cry from the cold little river I had crossed in St. Paul. The word "majestic" seldom really fits anything, except perhaps the world's great rivers as they near the sea; "majestic" is the only word to use to describe the Mississippi at Baton Rouge. It makes appealing what would otherwise be just another southern industrial town. There are a number of long barges and a number of big boats headed upriver, vessels bound for Memphis, St. Louis, Dubuque. Crossing this river at Baton Rouge, one can see why Mark Twain loved it and made it the subject of his richest and most graceful book, *Life on the Mississippi*.

Very soon after leaving Baton Rouge I came to one of my favorite stretches of the 10, the stretch of freeway that crosses the great Atchafalaya swamp—a land of mystery that extends much of the way between Baton Rouge and Lafayette. Long ago the swamp was beautifully filmed by Robert Flaherty in *Louisiana Story.* If "majestic" is a word that fits the Mississippi in its lower reaches, "spooky" is a good word for the Atchafalaya. Except for the bird sounds, it is largely a silent world—the snakes and alligators and other water creatures that inhabit it make little noise. Once in a while I spot a small boat moving through the dead gum trees. The swamp is a world that doesn't quickly yield up its secrets, although a Cajun culture still exists in and around it, in inlets, shacks, small towns. The only town of any size that's close to it, other than Baton Rouge, is Lafayette, probably the town most beloved by people who like south Louisiana, and its food, people, and music. For many Louisianans, New Orleans or even Baton Rouge is too large, too complicated, too big-city. For these folks Lafayette is their Paris. When they say the name they linger over the first syllable—Laaa . . . fayette—holding on to the word as if it were love itself. To them Lafayette means zydeco, jambalaya, crawfish étoufée, and lots of kicking up their heels on Saturday night or pretty much any night.

This wonderful area of swamp and silence is always somewhat under threat—despite the Army Corps of Engineers' elaborate strategies for controlling the lower Mississippi, most people who know and love the river know that it is more powerful than any plan human beings may design; one day it may rise up in flood and take out much of southern Louisiana, blowing through human constraints as easily as Moby Dick blew through the whaling boat.

Yet those who love the Mississippi choose to live by it anyway, seeing the river as a mainly benevolent god who yet may turn one day and become a destroyer.

AT LAFAYETTE I LEFT THE 10, the next few hundred miles of which would take me along the Petroleum Coast, where the presence of big oil is constant, both onshore and off. This area is best seen at night, when the belching and throbbing and wild flares of the great refineries give it an otherworldly beauty, a beauty that the light of day immediately destroys.

Through Lake Charles, Orange, Port Arthur, and Beaumont the presence of oil is palpable: you can smell it. The only part of this stretch of road that I like is the sudden opening onto the tidal plain that occurs just east of Houston; I like particularly the area around the Old and Lost River, where my character Muddy, in *Some Can Whistle,* was killed by a passing horse trailer while changing a tire on an overpass. The Old and Lost River gets lost, I suppose, in these tidal marshes—very soon after you cross it the skyscrapers of downtown Houston appear.

From Houston to L.A. the 10—for much of its existence as an interstate—has been my home road. I made the long drive west several times before there was a 10. Though President Eisenhower, in the grip of Cold War pessimism, conceived of the interstate system as a way of moving missiles and people quickly, much of the system didn't become a reality until the time of Lyndon Johnson. Before that time one traveled west or east in more or less continuous traffic, longing for an opportunity to pass the hay truck or groaning flivver just ahead. In the late sixties that began to change—drives that had once been hard became enjoyable; though even as late as the nineties, there were still big gaps in the system. The I-10 link through downtown Phoenix was only completed a few years ago; before that all east-west travelers had to slow down and submit to a four-and-a-half-mile crawl through suburban Phoenix.

The 10 is one of my favorite roads, but I haven't time, just now, to drive over stretches that I have already driven more than a dozen times. At Lafayette I turned north on the brand-new 49, an excellent short interstate that is really more inner than inter; it begins in Lafayette and ends in Shreveport, where it empties into the 20. It carries one pleasantly from swamp to scrub to suburb; peoplewise it carries one from the mostly gentle Cajuns to the far from gentle crackers of north Louisiana.

Although I was tiring by this time, I pressed on to Shreveport rather than spend a night in Alexandria, where I once had my worst-ever book signing. After having made a speech that was received at best sullenly I was trying to sign books in a dim hall, with no table and no help, when an old lady, very drunk, tried to swat me with her purse. "You didn't hardly write nothin' in my book," she said as I was trying to deal with other copies thrust at me by people I could barely see. I felt as if I had somehow stumbled into a pseudogenteel literary lynch mob, in a place where real lynch mobs were not an impossibility. Most of the people at my speech were drunk—in their greedy impatience they made a scene out of Hogarth, Gillray, Rowlandson. I have never so much as stopped for gas in Alexandria, Louisiana, again.

.   .   .

WHEN I ARRIVED back in Archer City at noon the next day I found that Jacksonville–Lafayette–Archer City had taken me almost exactly the same number of miles as Duluth–Oklahoma City–Archer City; there was less than ten miles difference. It was as if my love of organization had somehow prompted me to break the interstate system into eleven-hundred-mile chunks, perfect for a quick little run.

Eudora Welty said that trying to be a writer in Mississippi during the era of Faulkner was like living beneath a mountain. Although my route took me well south of Oxford, where he lived, I felt the presence of the mountain as I passed through coastal Mississippi. There were Faulkner paperbacks on the racks at the drugstore when I was in high school—there were even two books by his brother, John Faulkner (*Men Working* and *Dollar Cotton*)—but I didn't read Faulkner then. I read him in the sixties, in one long immersion, after I had come to know Malcolm Cowley, who helped put the then out-of-print writer back on the map by his inspired editing of *The Portable Faulkner.* Walter Benjamin said Proust was the Nile of language; in the same spirit one could call Faulkner—fittingly—its Mississippi; the Mississippi, at least, of the American language. There were lots of other good southern writers working during his lifetime, just as there were lots of other good French writers working during Proust's lifetime. The French writers may not have seen the mountain looming over them as clearly as Miss Welty saw the high peak just up the road, but the effect was much the same: two vital literatures were temporarily thrown into deep shade by the power of genius.

Faulkner himself, a little perversely, said he thought Thomas Wolfe was the best of his contemporaries because he tried to do the most; one suspects he made that judgment only for the plea-sure of ranking Hemingway fifth among his contemporaries. Wolfe didn't really try more than Faulkner; his characters tend to lose their edges, like jellyfish, in the sea of his words. Faulkner's power is tidal also, particularly in the extraordinary books of the late twenties and the thirties—*Light in August; Absalom, Absalom;* the great short stories—but his characters seldom lose their definition.

The other southern writer of genius—one whose gift was too distinct, too pure, and too original to be affected by Faulkner, was Flannery O'Connor, whose best stories are like nothing else in our fiction. She kept an eye on Faulkner, of course, remarking once that no one would want their old hack parked on the tracks when the Dixie Limited came whistling through. Faulkner was the Homer, the epic singer, O'Connor the silver poet, smaller but still very fine.

THE NIGHT I GOT HOME I sat for an hour with the road atlas, trying to choose my next interstate.

MARCH

The 70 from Baltimore to Burlington, Colorado. South on Highway 287

THE FACT THAT, for twenty-six or twenty-seven days each month, I lead an intense life as an antiquarian bookman—on the sorting floor all day, unboxing, pricing, sorting, and responding to the public's endless curiosity about *Lonesome Dove*—in part explains the brevity and intensity of my drives. I don't want to be gone from the bookshop long, but three or four days on the road, just looking and moving, isn't long. Working with books always relaxes me, but the books bring people, and people are a mixed bag; there comes a point at which I want to be away, drive somewhere, see some sky—it's my safeguard against the burnout that a month in the bookshop can occasionally produce.

Nagged a little by the sense that I ought to drive *some* road that's in the east, just for the sake of being representative or something, I hop a plane one Sunday afternoon and grab a rental car in Baltimore. For purposes of dealing with the east, the 70 out of Baltimore is ideal, mainly because it shoots one completely out of the east so quickly, particularly if one is starting one's trip, as I was, on a trafficless winter afternoon.

Also, I like Baltimore—despite its slicked-up touristy waterfront it remains a seedy old port city, reformable only to a negligible degree. It's the city of Poe and Mencken and, more recently, of John Barth and Anne Tyler, though perhaps it's wrong to include Barth, who's really the modernist of the eastern shore. Though he lives and teaches in Baltimore his writings aren't soaked in the life of that city in the way that Anne Tyler's are. Barth belongs more to the Chesapeake itself, to the great bay and the little crabbing communities that are strung along its eastern reaches. It's Anne Tyler who best captures the tired, insular weirdness of Baltimore. I had a tiny crack at it myself, in *Cadillac Jack,* situating the great flea marketer Benny the Ghost in Baltimore. When I was going to the city regularly, I did know someone a little like Benny the Ghost; he could be seen at any suburban swap meet or, in his urban mode, haunting the dusty aisles of the Harris Auctions on Lombard Street.

I descended into Baltimore on a cloudless day, the Chesapeake bright blue but flecked with white sailboats. It's only ten minutes from the airport to where the 70 begins. No sooner am I westbound than I'm reminded that these eastern roads carry one through a history-soaked terrain. A billboard lets me know that Frederick County, Maryland, is now 250 years old, making it nearly two and one half times as old as the county I live in, in Texas.

Very soon I'm in Appalachia, though at first only on its relatively prosperous eastern edge. At Frederick I am within fifteen miles of Waterford, Virginia, where my son and I lived from the late sixties to the early eighties. Waterford is an old Quaker village, set in a valley of picture-postcard beauty.

If Baltimore is the city of Poe, Mencken, and Tyler, it's also the city with a great hospital, Johns Hopkins. Some years back, as the result of heart surgery performed in the great hospital, I inadvertently left a self there—my first self, one that I was mostly comfortable with for fifty-five years.

Though it's a beautiful day for a drive, and I'm feeling just fine, I'm reminded of the odd way in which it came home to me that I'd misplaced that self as I pass by Hagerstown, Maryland. Though I had not quite put this together before, it was in Hagerstown, some weeks after my surgery, that I realized with bewilderment and sadness that a part of me—perhaps most of me—seemed to have died, been lost, vanished, slipped away.

The occasion that brought this sad insight was a trip, made in February of 1992, to look at some books belonging to the seer Jeane Dixon. The books, stored in an icy warehouse in Hagerstown, had been there for years—Miss Dixon was thinking of disposing of them. It was cold that day, and I was still a little weak from my operation, but neither cold nor weakness made that book trip different. From my earliest days as a book scout, or just as a reader, a chance to see books I hadn't seen has always excited me. My curiosity where books are concerned had always been intense—and has become intense again. But that day I didn't have it. I could see that Miss Dixon's books were interesting; they were mostly minor nineteenth-century books, but they *were* interesting—the problem, at first bewildering, later saddening, was that I just didn't care. After poking through perhaps a dozen boxes I went home. The curiosity that had always driven me to probe in box after box just wasn't there. I saw with my eyes that Miss Dixon's books were well worth having; but I no longer seemed to have the personality of a bookseller, which is to say, a personality that could have engaged with those books, box by box.

Somehow, without noticing it until that moment, I had drifted off—it was almost five years from that day in Hagerstown before I could really reconnect with antiquarian books or, really, any books. For two of those years I couldn't even read.

As I zipped along the 70 on a sunny afternoon, I could just see the warehouse in which Miss Dixon's books were stored—for all I know they are still there. It was in that warehouse that I had first noticed my absence from the affairs of living men—even from my own affairs as I had been accustomed to conducting them up to then. My curiosity had always had a natural link to my energy, and I had been blessed with plenty of both—then, for a time, I didn't have either. I stood in that warehouse feeling like a ghost, and continued to feel ghostly for some years.

Though it was quite warm the trees on the slopes of Appalachia as yet betrayed no hint of green; here and there, in the creases of the mountains, were fingers of snow, like little glaciers.

I soon left the farms of western Maryland behind, but not before noting that the planners who had jammed the 70 through this part of the country had made no effort to make farming look appealing. Invariably the freeway passed right by the mucky barnyards; travelers had only to glance sideways to see what shitty work farming can be. A little later, in the long valleys of western Pennsylvania, I saw some substantial farmhouses, the equal of any I had seen in Minnesota—some had as many as three silos.

I easily resisted the temptation to cut through Morgantown, West Virginia, where I could have picked up U.S. 40, the old National Road—the first federally funded highway, I believe. Too many times, in the long years before West Virginia had interstates at all, I had crawled painfully across the state, usually on mountain roads that offered no opportunities to pass the many underpowered coal trucks that would be groaning up the hills at two or three miles an hour, in their lowest gear. I remember my relief when I-64 was pushed through, making it possible to go from Virginia speedily over the hills to Charleston, West Virginia. The first time I traveled on the 64 the freeway was there and the exit signs along it had been put up, but no one had as yet got around to writing the names of the various towns on the signs. I was driving a big truckload of Oriental art-reference books to the University of Kansas, in Lawrence, Kansas. I had meant to stop for the night at the first town in West Virginia large enough to have a motel, but it was a foggy night and there were no town names on the signs at all. I soon lost all notion of where I might be and drove all the way to Charleston before I found a motel.

Resisting scenic West Virginia meant committing myself for a couple of hours to the Pennsylvania Turnpike in its western reaches. It's not what I would call a friendly freeway, being rough and pitted for most of its length, but I was glad to get to it because Pennsylvania, parsimoniously, has retained the fifty-five-mile-an-hour speed limit, even on some of its interstates. I was not to meet a hearty seventy-five-mile-an-hour limit again until I passed into

Colorado, more than a thousand miles west of the Pennsylvania Turnpike.

Along this turnpike there are still old brown barns with Mail Pouch tobacco signs painted on them—a form of advertising that, once those barns fall down, will cease to exist in America. Thanks to recent agreements, the Mail Pouch barns will soon be the only form of tobacco advertisements to be seen anywhere in America.

About all one can fairly say about the Pennsylvania Turnpike is that it will speed you east into Philadelphia or west to Pittsburgh. The western reaches of this highway take one into deep Appalachia. The region I was now in is rough, cold, hardworking—it produces great quarterbacks, that is, Joe Namath and Joe Montana.

It was almost dark when I crossed the Monongahela River, after which the road steadily worsened, as if, with only thirty miles of Pennsylvania left, the highway department had simply given up on it. Perhaps, too, the people along that stretch of highway had given up on themselves. I passed an adult video store whose neon sign was the brightest thing to be seen in all of western Pennsylvania. Or, perhaps, the occupants of the little narrow houses on the steep hillsides—houses built so close to the road that their occupants could have heard the radios, or even the conversation, from the passing cars—simply went on living their normal Appalachian lives, washing machines and old women on the porches, dogs under the porches, as if the highway that carried hundreds of thousands of strangers within thirty yards of their bedrooms didn't exist. The road was for strangers, but the hills were theirs, and the hills were enough.

On one or two occasions I've had car trouble in Appalachia and have clear memories of how difficult it is to interest the mountain folk in the troubles of outlanders. I once had a catastrophic break-down in southern West Virginia while traveling to Kansas City to make a speech. My only hope of getting to Kansas City on time was to persuade a local taxi driver in a little mountain town to take me to the airport in Roanoke, some forty miles away—for him an un-heard of adventure which could not be undertaken, despite all the money I was offering him, without prolonged consultation with his wife, who assumed that if he was going that far from home he must be leaving her. Finally, growing desperate, I suggested that his wife accompany us, which she did. Even then I'm not sure the man would have agreed to take me forty miles if I had not had my dog, Franklin, with me. Mountain reasoning seemed to be that a man traveling with a dog could not be all bad—even so, the second the man and his wife had deposited us on the curb at the airport they were off like a flash, headed back for their mountain fastness.

With that memory fresh in my mind I decided to stop for the night in Wheeling, an old town on the Ohio River, in that little snake's tongue of West Virginia that flicks up toward Lake Erie. The most convenient place to eat was a truck stop, with more than one hundred big rigs lined up in the parking lot. I try to avoid truck stops, the deep gloom of the truckers not being especially con-ducive to happy dining. Even those truckers who had a wife-companion traveling with them did not seem cheerful—couples and singles alike made do with lots of cigarettes and a loud jukebox.

The only West Virginia writer I can think of is Tom Kromer, author of a bleak little Depression novel called *Waiting for Noth-ing*, published by Knopf in 1935. It is not unlike Algren's first book, *Somebody in Boots*, published in the same year. Algren's book is dedicated "To Those Innumerable Thousands, the

Homeless Boys of America," despite which it is sometimes funny. *Waiting for Nothing,* never funny, is dedicated to "Jolene, Who Turned Off the Gas." Tom Kromer's father was a glassblower who died young; his grandfather had been crushed in a coal mine. What became of him after *Waiting for Nothing* I don't know.°

I had not driven far—just the later portion of a Sunday afternoon—and rose early the next day, meaning to go sunup to sundown, something that will be harder to manage as the spring days lengthen.

Accordingly I pulled across the Ohio River at Wheeling just as the rising sun glinted in my rearview mirror—thirteen hours later I watched it sink into the plains, beyond the Kansas River, at Lawrence. I spent the day in Ohio, Indiana, Illinois, Missouri, answering, at least to my own satisfaction, the question about where the midwest begins. Its eastern edge would seem to me to be Columbus, Ohio, after which the great cornfields appear and continue all the way to St. Louis. Eastern Ohio still has the look of Appalachia, with narrow frame houses on steep hillsides, but after Columbus there are no hillsides, just the flat midwestern plain. Columbus itself has grown a good crop of cornstalklike postmodern skyscrapers since last I was through it; it even produced a respectable traffic jam, though later in the day, at Kansas City, I was to see a really mature traffic jam, one that stretched for eleven miles; fortunately the mature traffic jam was producing road rage on the *other* side of the road.

In downtown Indianapolis the Chrysler plant had a big new

°I have since learned that the veteran *New Yorker* writer Philip Hamburger was born in Wheeling, and that the poet John Peale Bishop, Edmund Wilson's friend and sometime collaborator, was born in Charles Town.

blue sign saying "Daimler-Chrysler"; I had just read that at a big Daimler-Chrysler meeting in Geneva, the Daimler executives had addressed their new American colleagues in German, a brutal demonstration of Teutonic supremacy, some thought.

My first trip to Indianapolis, made some years ago, was to meet John Mellencamp, a native of the small town of Seymour, Indiana, some distance to the east. John was barefoot when he met me at the spiffy new Indianapolis airport, but he was driving his birthday present, a brand-new gray Jaguar. It was a little like being met by L'il Abner and taken to Dogpatch—only L'il Abner wouldn't have been driving a Jag and wouldn't have owned Dogpatch to quite the extent that John Mellencamp seemed to own Seymour. The projected movie that had brought me to Seymour eventually got made as *Falling from Grace;* besides which, John Mellencamp produced my son's first two record albums, though I suspect that he still seldom bothers to wear shoes when meeting folks at the airport, the Daisy Mae that could tame him having perhaps not yet appeared.

An odd fact that pops into my mind as I rip through Indianapolis is that Janet Flanner grew up there before moving to Europe, where, writing as Genêt, from 1925 until 1975 she delivered her brilliant Paris letters to the *New Yorker:* fifty years of high-level journalism that would seem to have nothing midwestern in it except Janet Flanner's solid common sense. In the pantheon of good *New Yorker* writers she should stand higher than she does; she was as good as A. J. Liebling or Joseph Mitchell and lasted quite a bit longer; those letters read as well today as they did when they were published.

For much of this day and most of the next, from western Pennsylvania all the way to Colorado, I-70 runs just parallel to or actually encompasses its parent road, old U.S. 40, the so-called National Road, subject of a good book by the now much underappreciated writer George R. Stewart. In fact, the *only* appreciation of George Stewart that I know is in a late book of essays by Wallace Stegner, *Where the Bluebird Sings to the Lemonade Springs*. Stewart, among other skills, was an authority on American place-names—he compiled what is still the best dictionary of them. His study of U.S. 40—the first federally funded interstate—is leisurely and informative. The first time I struggled over the Rockies I did it on U.S. 40, which nowadays merges with I-80 at Salt Lake City. George R. Stewart, like Walter Havighurst, is one of a fairly large number of American writers who rose, mostly from newspaper work or college teaching, to make admirable if little noticed careers, writing honestly if rarely beautifully about the subjects they took on, leaving a valuable record. I still go back to them when I need information about how certain places were at certain times. Like Janet Flanner, of Indianapolis and Paris, they were solid.

I think I have a pretty high curiosity, but there's really not much for even the highest curiosity to latch on to as one follows the 70 across Indiana and southern Illinois. In the early nineteenth century, in the tiny little community of Fall Creek in Indiana, a group of locals massacred a small band of peaceful berry-picking Indians; so far as the locals knew, killing Indians was not a crime. But the U.S. government, fearing that the great tribes of the old northwest would rise up and wipe out the frontier, made Indian killing a crime retroactively, tried the men, and hung five of them. The Indians, in their robes of state, came to the show trial and accepted the verdict. Jessamyn West wrote a workmanlike novel about this incident, called *The Massacre at Fall Creek*. The English director Jack Clayton and I worked amiably on a script about this singular incident, but no movie was ever made.

I wanted to get through St. Louis before the afternoon rush hour, and just made it, though my efforts to study the Gateway Arch were frustrated by this need for speed. But then, slow or fast, it is frustrating to attempt to admire the Gateway Arch because of where it is placed. To get the right kind of look at it one would need to be atop one of the buildings in downtown St. Louis, or perhaps in a helicopter. Studying it from the park underneath it is no good. One needs to be able to see *through* it to the great west beyond. The arch, Eero Saarinen's masterpiece, is, in my view, probably the most beautiful manmade structure west of the Mississippi—had it been placed on a lonely bluff somewhere up- or downriver, it would have been as astonishing as anything in America. But as it is, if you look east through it you just see the squalor of East St. Louis, and if you look west there's the mass of downtown St. Louis. You cannot look through it and see the west, despite which the Gateway Arch remains a great thing. In my mind's eye, when I drive past it, I keep transferring it to a lonely bluff upriver; to allow the eye to be tempted as one negotiates the fast, mean interchanges of downtown St. Louis is to invite instant smashup. That the Gateway Arch has not been given a context commensurate with its beauty is a great pity. It should invite the eye into the new country, rather than just to the confluence of the 70, the 55, the 64, and so forth.

My attitude toward Missouri, already expressed in my comments on I-35, is that it's a place to get through as rapidly as possible. Once one clears St. Louis, if one is going west, this means being in Missouri about three hours, which is certainly enough for me. The roadsides offer few attractions, all of them low. I remember being shocked, when I was delivering that load of Oriental art books to the University of Kansas, by what came on the TV when I stopped for the night in Independence, hometown of the staunchly upright Harry S. Truman. What came on the TV was pornography, which I had certainly not expected to encounter in a motel only a few blocks from the Truman Library. That delivery was made in the early eighties, when one would not have expected to encounter pornography anywhere, except possibly on Forty-second Street or Hollywood Boulevard. Yet there it was, in Independence: a man who seemed to have extended his already considerable organ with what appeared to be a length of garden hose was inviting a woman to do stimulating things to it. I had driven an underpowered truck a long way that day and was too tired to be much titillated, but I have not forgotten that what has become standard fare in motels all over the country first swam into my ken in Missouri.

The next day I had Kansas to cross, a slightly longer toot than the Florida panhandle. It's not necessary to progress far in Kansas to notice how adept the roadside communities have become at hustling their history, of which there is quite a bunch. I'm hardly out of Lawrence before I begin to see billboards advertising the Holy Triumvirate: not Father, Son, and Holy Ghost but Custer, Cody, and Hickok, all of whom did have notable adventures in Kansas.

Manhattan, Kansas, in a burst of wit, has decided to call itself the Little Apple—it offers visitors a number of attractions (including the last existing patch of tallgrass prairie), but not nearly so many as its neighbor Abilene boasts. Abilene, Kansas, one of the first places to achieve notoriety as a cow town when the great herds were coming up the trail, offers visitors such a dazzling variety of museums and halls of fame that I had to stop for a moment, to scribble down a few of the most eminent. There's the Dwight D. Eisenhower Museum and Library, of course, but there's also the Museum of Independent Telephony, the Dickinson County Historical Museum, the Kansas Sports Hall of Fame, the Greyhound Hall of Fame, a Plains Indian Museum, and plenty of places in which one could eat in a historic setting.

Earlier in the morning I had stopped and poked around a little at Council Grove, in the nineteenth century a significant jumping-off place for adventurers, traders, or emigrants headed west on the Santa Fe Trail. Josiah Gregg has a good description of it in his *Commerce of the Prairies.* From St. Louis to Council Grove—with the Missouri River to follow for much of the way—the trip west was comparatively easy and comparatively safe; but beyond Council Grove the great, dangerous prairies yawned. When Charles Bent and Ceran St. Vrain made their first effort to establish a trading route from St. Louis to Santa Fe, young William Bent was allowed to go only as far as Council Grove before being sent back in bitterness to St. Louis, to manage the store, as it were. From Council Grove west there were no guarantees—many a hopeful emigrant didn't get much farther.

As one proceeds west on these prairies the 70 can be seen to have taken a bad pounding from the great trucks. Travelers, whether they want to or not, can recapture the feeling of what it was like to follow U.S. 40 when it was a two-lane road. I didn't want to recapture that feeling—I had inched across Kansas often enough on the old 40—but I had no choice, since more than sixty miles of the route was under heavy construction.

In western Kansas the local offerings, in the way of museums and exhibits, get rapidly slimmer. One small community could come up with nothing better than a five-legged cow, while another tempts tourists with a chance to inspect the world's largest prairie dog—whether this monstrous creature is alive or stuffed isn't specified. The small town of Victoria, Kansas, named, I suspect, for the old queen, offers travelers a look at the Cathedral of the Plains. I took a look—I don't know that the church is exactly a cathedral, but as a piece of American Gothic it's well worth the stop.

Throughout western Kansas the plain is immense, the horizons a very distant blur. I love such country and sailed through it happily. The only trouble seemed to be that my clipper ship, again a sensible Buick, seemed to be listing strongly to the right; the reason for this listing became obvious when I stopped for gas in Russell, Kansas, Bob Dole's hometown. The cause of the drift was wind, a wind coming from the south with such force that it pushed me past the door of the gas station. I discover, in the course of this stop, that Senator Dole and I have something in common, namely highway 281, which runs right through Russell and on past my ranch house, many miles south.

As I push on toward Colorado the wind increases—most farmhouses in this region are protected from such winds by thick growths of cedar trees, planted as close together as possible; rarely are the houses themselves visible through the trees since, if the eye can get through, so can the wind.

When I crossed the Colorado line and stopped for a sandwich

in Burlington I glanced at my speedometer and discovered that I was precisely fifteen hundred miles from my starting point in Baltimore. Fifteen hundred miles is enough of the 70, at least for now. If I keep going west I'll only have to contend with Denver—and I'd rather not. Until the boom of the late sixties, Denver was a jewel, perhaps the most appealing small city in the west. Jack Kerouac rhapsodized it as it was then, in *On the Road*. As the hometown of Neal Cassady, Denver has a solid place in Beat mythology—see Allen Ginsberg's "Green Automobile," his ode to Cassady. Once the Denver air was tonic, but both the air and the ambiance suffered as Denver prospered. Its new airport, with its dazzling white tent domes, can be a little disorienting when one comes to it from the plains to the east—is that Denver we're approaching, or is it Mecca?

At Burlington I turned south and traveled all afternoon just inside the Colorado line, through the cities of Cheyenne Wells, Kit Carson, Eads, Wiley, Lamar, Springfield, Campo—I came into the Oklahoma panhandle just above Boise City. Despite the wind, which blew in the relentless way which tormented and not infrequently destabilized pioneer women, who had come to that plain from sheltered places, I had a beautiful drive, just sky and plain to look at, the prairies stretching on and on.

Part of my interest in looking at eastern Colorado had to do with the Bent family, traders who were the subject of a good if largely forgotten book called *Bent's Fort*, by David Lavender. The Bents were a diverse, tenacious family who, with their French partner, Ceran St. Vrain, were for a time the principal traders on the Santa Fe Trail. Their first trading post, or fort, was on the Purgatoire River, a tributary of the Arkansas. This fort was blown up in 1849 by an indignant William Bent—after the Mexican War, when New Mexico suddenly became America, William Bent tried to sell the fort to the army, but was offered such a pittance that he blew the place up rather than let the army use it. But he soon built another fort, on the Arkansas itself, and then another, far south on the Canadian, which became known as Adobe Walls. This one was in the dangerous country of the Comanche and the Kiowa—it was to be the site of two famous battles. Of course, all the west was dangerous country when the Bents began to trade in it, around 1830. Charles Bent, the eldest brother, was hacked to death in the Taos uprising of 1847; his brother Robert was killed by the Comanches. William Bent, who did most of the managing of what was once a considerable trading empire, liked the Cheyenne people and married two Cheyenne women, first Owl Woman and then her sister Yellow Woman, who was killed by the Pawnees. The métis sons of these unions mostly preferred their mother's people to their father's. One of William's sons was forced at gunpoint to lead John Chivington, the Fighting Parson, on November 29, 1864, to the village of the peace-loving Cheyenne leader Black Kettle. Two of the young man's métis brothers were in the village at the time, visiting their Cheyenne cousins; they narrowly escaped death at the famous massacre that followed: Sand Creek.

My southerly route down eastern Colorado goes near the site of this massacre, and near is close enough for me. The mutilations performed on the Cheyenne victims, men and women, got much attention in the war-weary east, encouraging the Peace Party to cry out (for a time) about justice for the red man; but, in kind, that massacre, that terror, was no different from hundreds of other acts of massacre that had occurred since Europeans and Native Americans first came into conflict on this continent. Sand Creek just got more press. One of the saddest elements, for me, is that Black Kettle, who survived the massacre, could never get the whites to recognize that he was a peace Indian, and had been from first to last; but many white men never cared to distinguish between friendlies and hostiles when they were out to punish Indians. Four years later, less a week, the Son of the Morning Star himself, George Armstrong Custer, led a famous dawn attack on Black Kettle's village, then well to the south, on the Washita. This time Black Kettle was not so lucky: he and his wife were both killed.

By an unhappy but amusing accident, on my way south I missed my chance to inspect the one historical Point of Interest along my route—though I have since repaired to my guidebooks, I am still not sure what I missed.

As I was proceeding west, between Cheyenne Wells and Kit Carson—a town named for the famous and durable scout, who was, it's startling to reflect, only about an inch taller than my very short agent, Irving Lazar—I had the misfortune to get behind a young farmer who was pulling a very wide plow. This expensive piece of equipment was actually wider than the road we were traveling on, and as the bar ditches on both sides were deep and steep, there was no way the young man could pull over and let me go by. He was traveling about ten miles an hour, and it was some twenty-five miles over to Kit Carson. I was not bothered by this slowdown—I was happy enough just to be looking at the plains.

But the young man, conscious that he was holding up traffic—one car—*was* bothered. He glanced back several times to see whether I was fuming with impatience; I tried to reassure him but could only do this by waving in a friendly fashion. Just then the solution to his dilemma appeared, in the form of the Point of Interest, which provided a wide turnout. The young farmer with the plow turned out and parked directly in front of the historical marker itself, which showed a cavalryman on a horse. I could have turned out too, and waited until the young farmer had gone on, but that would only have perpetuated our problem; the marker was not likely to occupy me for more than thirty seconds, after which I would be behind him again. Besides, the young farmer was obviously pleased with his courteous solution to the problem of the small road and the large plow.

So, figuring I could find out who the cavalryman was from the guidebooks, I waved and swept on by, only to find that the guidebooks held no definitive answer—at least the ones I own don't. It is not likely that Colorado would have put up a statue to Chivington, who, in any case, would have been coming from the other direction; nor is it too likely to be General Hancock, who led an expensive, entirely futile expedition across these plains in 1866, meaning to punish the southern Cheyenne, whom he never found.

My best guess is that the cavalryman on that pedestal may have been Stephen Long, who led an expedition across these plains in 1820. I think so because the statue is near the now defunct town of Firstview, the name coming from the fact that once, around about there, on a clear—very clear—day, it was possible to get a first glimpse of the Rockies as they rose from the broad plain. I don't think it's likely that there will ever be a day that clear again, what with all the pollutants that now lie between the traveler and the mountains, but there must have been such days once, because a young lieutenant named Samuel Seymour did a quite appealing painting of his own first glimpse of the Rockies, not far from the place at which the young farmer pulled off to let me pass.°

Anyway, I passed on south, mildly mystified. Every driver I met waved to me, except one who was talking on a cell phone—it's a lonely, long road, though it became less lonely when I came to the area of big agribusiness, whose headquarters town is Lamar, Colorado. What had been untended prairie became irrigated fields, some with irrigation trestles as long as three hundred yards stretching across them; this region of irrigated farmland extends all the way south to Vernon, Texas, which is only sixty miles from my home. All this water comes from the great Ogallala aquifer, the liquid resource that has transformed the southern plains. What will happen to the vast fields once this aquifer dries up is not a topic anyone on the plains wants to talk about much; a whole region has become dependent on an exhaustible resource—underground water—the result being that the boom psychology that is so common in the oil business has somehow transferred itself to the farmers. They know the water can't last forever, but they probably hope that it will last as long as they last.

By the time I pass through Lamar the soil from all these plowed fields is coming at me on the wind; I began to feel rather like I feel if I happen to be in the vicinity of Lubbock, Texas, when the wind is blowing.

At Lamar I cross the Arkansas River—the Bents' second trading post is not far upstream. I am now going south on highway 287, a road that runs all the way through Texas to the Gulf at Cor-

°Lieutenant Seymour's painting is reproduced in *The West of the Imagination,* by the Goetzmanns, *père et fils,* page 12.

pus Christi. In Colorado the 287 is mainly a road for locals or professionals, by which I mean truck drivers. All afternoon I meet herds of trucks hurtling north, sped on by the powerful tailwind.

The Four Corners area that most people think of when they hear the term is the conjunction of northeastern Arizona and southeastern Utah with northwestern New Mexico and southwestern Colorado—the area where the murderous survivalists slipped away from a vast pursuit in the summer of 1998 (they are still unapprehended). But as I pass through the Comanche National Grassland I am in another region where the corners of four states lie close to each other—Kansas, Colorado, New Mexico, and Oklahoma—though here the corners are not jointed quite so neatly as they are to the west.

The Oklahoma panhandle, which I cross in forty-five minutes, is still as lonely, lovely, and spooky as ever. There is some farming, but it is still mainly an area of ranches, inhabited by hardy souls who like to be left alone to fend for themselves. The fact that their nearest neighbor might be thirty or forty miles away not only doesn't bother them; it's the reason they're where they are.

I finally break back into Texas just north of Stratford, a small town very unlike the hometown of the Bard of Avon. One geographical fact I can never seem to keep in mind is that more than one hundred miles of Texas territory lies north of Amarillo. In my mind's eye I keep putting Amarillo at the top of the state, when it isn't even very close to the top. Grinding down to it, through the last of the day, is no pleasure, though I am cheered, as I'm about to drop off into the breaks of the Canadian River, to see a small sign on my left that says "Bent Ranch." The site of the Bent brothers' old fort on the Canadian—Adobe Walls—is not far away. Since so many of the Bents, in both the first and the second generations, met violent ends, it's nice to know some of their descendants are still there, still part of the life of the high plains which Charles and William Bent crossed with such hope in 1830.

I spent the night in a motel in Amarillo, on the littoral, or near shore, of interstate 40, a great road, surely, but one I know too well and don't much like. All night the trucks scream through Amarillo, a herd indifferent to night or day. The high whine of tires on concrete never ceases—like the sigh of the wind, the whine of tires on concrete is something one gets used to if one's pleasure is to travel the great roads.

The next morning I proceed south down the 287, meaning to be home before my bookshop opens, even though this means plunging a little sleepily through an area fraught with personal significance.

The first town I come to, while it is not yet full light, is Claude, where, in 1961, Hollywood entered my life. *Hud,* the movie made from my first novel, was filmed in and about Claude. I was far down the road at the time, teaching at TCU, in Fort Worth. One day a nice man from Paramount whose job it was to find and arrange for movie locations showed up in Fort Worth and took me to dinner, to pick my brain about possible locations. I gave him the name of my cousin Alfred, then a resident of Clarendon, a few miles down the road from Claude. Alfred soon rented the movie company some of his ranch, many of his cattle, and even a few of his cowboys. Shortly after the nice man's visit, Paul Newman, Patricia Neal, Brandon De Wilde, and Melvyn Douglas were all bivouacked in Amarillo, filming *Hud.* Since I was still teaching school in Fort Worth I visited the set only for one day, a visit which I later described in *In a Narrow Grave.* The high point for me was just passing through Claude and noticing that they had changed the name on the water tower to Thalia, the name of the fictional town in my book. (The real Thalia, Texas, just a shipping point on the railroad that once boasted a filling station and a small store, has since been thrust into prominence—it was for a time the hometown of Kenneth Starr.)

The next town of consequence down the 287 was Clarendon, which, for a short time just after World War I, was the seat of the McMurtry family. The nine McMurtry boys and the three girls fanned out from it, acquiring wives or husbands as well as ranches and farms. I found out quite by accident recently that my father actually graduated from the then-tiny Clarendon College, class of 1921. But the old folks didn't like the high plains and went back to Archer County—graduate or not, my father went with them. All this I have written about in the final essay of *In a Narrow Grave.* My father is gone and Clarendon College is no longer tiny. I don't know how great a challenge it presents academically, but situated as it is on a high windy ridge, the physical challenge of getting from building to building when the wind is blowing cannot be inconsiderable.

South of Clarendon about the only fun to be had is crossing the Red River, as it meanders in a shallow stream between wide banks liberally covered with salt cedars. The Red still holds a certain romance for me, perhaps because in earliest youth I heard my uncles—several of whom ranched along it—talk of cattle or even horses lost beyond hope in its treacherous quicksands. Crossing the Red River was always an adventure for them, and their descriptions of it were, for me, a source of permanent awe.

It doesn't really look like a very dangerous river—nothing like the swift, narrow-banked Pecos—but when I cross it in my mind's eye I still see those abandoned cattle and horses sinking out of sight in the long, reddish stretches of sand.

I passed through Claude, Clarendon, Memphis, Estelline, and Childress, slowly cruising each main street in hopes of spotting a cheery cafe in which to breakfast, but no cafe that looked the least cheery could be found. In Quanah, too hungry to wait any longer for cheer, I stopped at a very uncheerful-looking drive-in, only to find six men sitting at a long table laughing heartily at a joke one had just told. One of the men recognized me and told me he liked my writing. I was exactly one hundred miles from home, about the outer limit of my area of recognizability. I pulled into Archer City just as the girls were opening the bookshop.

TWO WEEKS LATER, in mid-April, I composed a kind of coda to this drive. I had promised to help dedicate a new public library in Pampa, a high plains town about thirty miles north of I-40, so I wound back up the 287, a road now boresome to me, and then crossed east of Amarillo to Pampa, where I did my job and visited for a bit with Timothy Dwight Hobart, an old friend whose grandfather had surveyed much of the Texas panhandle.

Tim showed me the sights of Pampa—they took not long to see and included a house where Lawrence McMurtry, the uncle I was named for, had lived. Lawrence McMurtry fell to his death from a grain elevator sometime in the twenties—my father, who was working for him then, picked him up and rushed him to a hospital, but to no avail. Lawrence is the McMurtry I know least about. It is rumored that he drank, and may have been drinking when he fell out of the elevator, but—at a remove of seventy years—this is merely legend.

That morning, inasmuch as I was early for my speech, I drifted east a little distance on the 40. My drift took me as far as Shamrock, Texas, where my aunt Margaret had lived most of her short and not notably happy life. In my boyhood she had been my chum, the only one of the twelve McMurtrys companionable enough to be a little boy's chum. Sadly, she had no children of her own. We went skunk hunting together, wandering over the family hill with a .22 but damaging no skunks. Her sister Grace, eldest of the three McMurtry sisters, lived in the neighboring town of McLean—her I scarcely knew. Indeed, I scarcely knew any of my father's eleven siblings—just Aunt Margaret, and all I knew about her was that she was sad.

The first McMurtry family reunion was held in 1918, on Charlie McMurtry's ranch, in Donley County, near Clarendon. I have a photograph made on this occasion, in which the family in its pride is standing together, posed against a stark roll of the prairie. There are some forty people in that photograph: my grandparents William Jefferson and Louisa Francis, their twelve children, and the children's wives and numerous progeny. I've studied that picture often, wondering about the lives of all those strong and striking people, my kinsmen and kinswomen, there portrayed. But the stark and treeless prairie behind them is the dominant element in the photograph, not the forty McMurtrys. Here on the fertile high plains the McMurtrys had made their stand, and it was a vigorous stand, for they were a vigorous breed, all of them endowed with a huge capacity for work and, as is evident, at least a healthy tendency to breed. But what, really, could their strength do against that beautiful, sere, remorseless landscape?

Returning south from Pampa on a little prairie road, I passed within about three miles of the spot where that picture was taken, some eighty-one years ago. The plains are still there—along that stretch of road they are almost unchanged: still no trees, just the waving grass and the rolling land—but those McMurtrys are gone, all of them—even the youngest baby in that picture is now dead. The years have taken them; for all their vigor, there is not much to mark their passage. I made my speech where, as recently as the 1950s, there had been at least half a hundred McMurtrys in action, performing their civic duties, such as the one I had just performed. But all flesh is grass, as the old saying goes: the grass is there still but the McMurtrys are gone. One old cowboy shambled up after my speech and told me that my uncle Charlie had raised him. A woman appeared who was related to my uncle Roy, I didn't quite catch how. And that was it. I was speaking only three blocks from where my uncle Lawrence, drunk or not drunk, had fallen to his death, and yet only three people in that crowd seemed to recollect that there had been a Lawrence McMurtry— that he was remembered at all was mainly because of his widow, my aunt Gertrude, who survived him by some years.

Along the 40, east of Pampa, a fragment of old route 66 has been rather studiously preserved. For the sake of history, not nostalgia, I drove along it for three or four miles. My hero Captain Randolph Marcy, with George B. McClellan accompanying him, anticipated both the old 66 and the present I-40 when he passed along that very stretch of prairie in 1852, attempting to determine which fork—north or south—of the Red River was the true channel, a question that was decided in favor of the south fork, which proved convenient much later when the south fork of the Red became the border between Texas and Oklahoma.

On my way home from dedicating the library I saw, just south of Electra, a herd of camels grazing near the road. This supports my thesis that one is apt to see anything along the American road. Once in the Sierra Nevada, in the middle of the night, I even saw two elephants crossing the highway; they had just escaped from a circus and remained free for several days.

But the camel herd south of Electra had been there since my boyhood—perhaps seeing them in earlier years is what prompted me to put the camels into the Uncle Laredo chapters of *All My Friends Are Going to Be Strangers*. I rarely drive the road between Electra and Archer City and tend to forget about those camels. I may not have noticed that camel herd in something like forty years. Even though it soon came back to me that this was a long-established camel herd, and that I had known that they were there for many years, the sight of them still gave me a start. I felt, for a second or two, when I saw the ungainly animals slipping through the mesquite, that reality had just slipped out of synch. I felt somehow in the wrong place: how did I come to be where camels were? The sight produced a momentary, but powerful, sense of dislocation.

Sights along the American highways have often given me that brief, sharp sense of dislocation. On my previous drive south from the 70, another such moment had occurred. As I was smoking along the 287, just south of Dumas, Texas, I suddenly saw something immobile right in the center of my lane, some distance ahead. It looked like a swivel chair, such as the boss of a little oil company might sit in when in his office.

The chair wasn't upended, either, as one would suppose it might have been had it fallen out of a truck. It wasn't on its side. It was a well-padded, brand-new swivel chair, sitting right in the middle of my lane, waiting for the boss to come sit in it and start making calls on his cell phone. Fortunately the 287 was four-lane at that point; I was speeding along at about eighty-five miles an hour and was only just able to swerve around the chair, once I accepted the fact that there had been a chair in my lane. There was no traffic behind me or ahead of me at that moment—there was just the chair, and myself in a car, and the great plains on either side. I was past the chair in a moment, but it was an eerie moment.

Perhaps two miles later I saw the pickup the chair had fallen out of, frantically backing toward the dot in the distance that the chair had become. Until I saw the pickup, with three other identical swivel chairs in it, I hadn't entirely believed what I had just seen. I had driven all the way from Lawrence, Kansas, that day and was hurrying along; I swerved past the chair before my brain could quite process the message my eyes had just given me. *Was there a chair in the road, or had I just been daydreaming?* The high plains have long been famous for the mirages they create—some think it was these mirages that drew Coronado onward; he was seeing golden cities where there was really only sky and plain. Perhaps, Coronadolike, I had imagined that swivel chair.

But then I saw the pickup, backing. The backing was itself an illogicality; it would have been both quicker and safer for the driver to have turned around and zoomed back up the highway to where the chair sat. The pickup was veering this way and that as it backed. I realized, with a slight pang of guilt, that there really had been a chair and that I should have stopped and removed it from the roadway before some eighteen-wheeler hit it and reduced it to kindling.

Reflex had allowed me to swerve around the chair before my brain fully assessed its presence, a surreal, Magrittelike presence, all told; by more or less the same token, memory soon enabled me to come to terms with the camels, which had been grazing in that pasture for most of my life. And yet surreality will often be a part of experience on the road, particularly if the road crosses the plains. Mary Ellen Goodnight, the wife of the great cattleman Charles Goodnight, once mistook a clump of sage grass for a party of warring Comanches; she even convinced the cowboys who were with her that the sage grass was really a large group of Indians. In their anxiety about Indian attack, the cowboys held up the progress of a cattle herd for most of a day, while they studied the situation. When Charles Goodnight returned, he glanced, with his eagle vision, at the distant clump of grass and pronounced it to be grass, not Indians. He was not happy about the fact that neither his wife nor his cowhands could tell grass from Comanches. Yet in his many years on the prairies, Charles Goodnight himself was well acquainted with their power to create mirage; on his many drives he had often seen inviting lakes of water shimmering in the distance, where no water was. Many nomadic peoples, in their wanderings, often feel that they see things that turn out not to be there—things, in many cases, just as strange as a swivel chair in the road.

APRIL

The 75 from Detroit to the Sault Ste. Marie.
Michigan Highway 28, the 39, the 90–94, the 29

MY PLAN FOR APRIL, conceived somewhat reluctantly, was to take I-75 all the way south from the Sault Ste. Marie to the Everglades Parkway, after which I would fly home either from Tampa or from Orlando, depending on my whim. My reluctance was due to the fact that I find the middle reaches of the 75 pretty boring, and its middle reaches might be said to extend all the way from Toledo, Ohio, to Valdosta, Georgia, or thereabouts. Kentucky and Tennessee are pleasant enough states to drive across, but Ohio and Georgia really aren't pleasant; besides, I had just done an east-west crossing of Ohio and did not feel any immediate urge to traverse the state north to south.

Still, I was prepared to put up with the dull midsection of the 75 for the sake of its northern and southern tips. I wanted to go way up in Michigan and way down in Florida, perhaps as far as the Keys if I was still in a driving mood when I got to the Everglades.

Accordingly I secured a rental car in Detroit and was just about to head up to the Sault Ste. Marie when I heard the word "Everglades" and glanced at a TV in the airport bar to see a wall of flame. The Everglades, it seemed, had caught fire; sixty miles of the 75 below Tampa had been closed. No rain was predicted, either; the Everglades were going to burn for some time, and very likely would continue to smoke and smolder for a month or two once the flames were extinguished.

This was discouraging—the very part of the road that I would be looking forward to all the way down through Georgia and north Florida had just been put off limits. One thing I don't seek is smoke. The summer of 1998 was mostly ruined for residents of Texas and several other states by an acrid, smoky haze that had drifted north from some ten thousand agricultural fires in Guatemala and southern Mexico. I had hated that haze, and saw no point in driving all the way south from the Great Lakes to experience something similar again.

Surprises, though, are part of travel, and there were lots of other routes I could pursue once I had seen northern Michigan, something I was determined to do. I meant to go all the way up to the St. Marys River, in homage to two American writers who had written beautifully about the Upper Peninsula: Ernest Hemingway and Janet Lewis. As a rule I am not much inclined to pilgrimage, literary or otherwise, but these two writers had made northern Michigan so vivid in their writings that I wanted to see the place for myself. Although both of them wrote of northern Michigan as it was long ago, it seemed likely that, because of its relative remoteness, the Upper Peninsula would be country that hadn't changed very much.

The year before, I had written an essay about Janet Lewis and had flown out to California to have supper with her and listen to her talk about some of the writers she had known, one of them being Ernest Hemingway, with whom she had been in high school. At the time I met her, Janet Lewis, ninety-eight years of age, was still living in the same house, near Stanford University, that she had moved into with her husband, Yvor Winters, in the 1930s. She died less than a year after my visit. Though her vision was a little cloudy she was still in full possession of her faculties and talked at length about Hemingway; his sister Marcelline; her long, happy marriage to Yvor Winters; Hart Crane; and northern Michigan as it had been when she was a girl. Janet even admitted that Vladimir Nabokov had liked her so much that she had, from time to time, allowed him to help her wash dishes. Her own achievements in both poetry and fiction are distinctive, the short novel *The Wife of Martin Guerre* being, in my view, her master-piece. But it was her vivid first novel, *The Invasion,* that drew me to the north country. Though the novel is essentially the lightly fictionized history of the well-known and historically important Johnston family, it is Janet Lewis's confident, abundant descriptions of the northern forests and the great waterways which allowed settlers such as the Johnstons to penetrate them that make *The Invasion* a compelling book. From first to last Janet Lewis wrote wonderful prose, very different from that of her famous classmate Ernest Hemingway. The rich shadings and long cadences she uses to describe the forests in their various seasons bring to mind the thick prose Faulkner uses to describe the big woods of Mississippi. I wouldn't push that comparison too far, though I do think landscape partially determines the prose writers use to describe it; dense forests seem to prompt writers to a corresponding density of expression.

Hemingway's reputation is now so encrusted with a seventy-five-year accumulation of biographical lore—most of it unfavorable to him—that it is easy to forget how well he wrote at the beginning of his career. Picasso said somewhere that no one had looked at Matisse as hard as he had. Hemingway, as a young man, had the same ability to look hard, as hard as a painter, at a place and at the often destructive human emotions operating within a place that was seemingly as lovely as northern Michigan.

Hemingway left his Michigan stories scattered, but Philip Young pulled together a book of them, called *The Nick Adams Stories*, putting them into a kind of sequence. Hemingway's instinct—to leave them scattered—may have been wiser. They are, to some degree, about isolation, and they strike with more force when encountered in isolation, here and there in the early story collections.

Many who came to dislike Hemingway forgot how good he was; he may have forgotten it himself, or he may have just been unfortunate, over the long haul, to have perfected a style that had nowhere to take him but into self-parody, which is where, as a writer, he resided for so many sad years.

It was because of those two writers, Janet and Ernie (as Janet called him), that I found myself wending my way north through the western suburbs of Detroit on a gloomy, overcast day in April.

Because the airport is placed well to the southwest, I didn't have to engage much with Detroit itself—I simply went north through Dearborn, which is still very much the land of Ford. One of the first large structures I passed was a Ford plant, followed shortly by the Henry Ford Museum. Henry Ford thought history was bunk, but then, through genius, became part of it, so much so that he has a museum named for him.

It is not necessary to travel far around the populous south shores of any of the Great Lakes to feel oneself in the land of Big Industry. The warehouses and shipping sheds that border the big lakes are as immense as those I saw north of Laredo, at the end of the 35.

On this drive, though, I don't really see any water until I pass Saginaw and come briefly in sight of Saginaw Bay. Not far north of Bay City the 75 turns decisively inland, splitting the last two hundred miles of the rather fat peninsula right down the middle. For most of the way north I'm in farming country—evidently prosperous farming country, judging from the size of the houses. The country is quite flat, as flat as the Minnesota farm country south of Lake Superior.

The only visual excitement on my drive north is provided by the splendid arch of the Mackinac Bridge, a toll bridge which takes one high above the Straits of Mackinac, the narrow waterway connecting Lake Michigan with Lake Huron. Just on the south shore is the old Fort Michilimackinac, where many of the struggles that Janet Lewis describes in *The Invasion* took place. Those contesting the area were Ojibwa (or Ojibway), French, British, Americans, adventurers, soldiers, trappers, priests. Though the Ojibwa were a populous people whose territory stretched as far west as the Turtle Mountains in North Dakota, they were never politically unified, in the manner of the Iroquois, and offered less resistance to the complex wave of Anglo-French invasion described in Janet's book.

Just before crossing the Mackinac Bridge I at last succumb to a "museum"—in fact, only a store pretending to be a museum, called Sea Shell City. All the way up from Saginaw I have been seeing billboards describing the beauty and delicacy of the seashells on exhibit in Sea Shell City—but then, at the very last billboard, as if sensing that beauty and delicacy may not be enough to slow the bloody-minded hunters heading north, the sign promises a five-hundred-pound man-killing clam, a traffic stopper for sure. Though, in Kansas, I had easily resisted the five-legged cow and the world's largest prairie dog, I found that, jaded as I am when it comes to roadside attractions, I could not resist a man-killing clam. I whipped into Sea Shell City, only to discover that I had arrived too late; the store had just closed. Peer though I might through the big windows, I could not spot a five-hundred-pound clam.

Once across the Mackinac Bridge it is not far to the top of Michigan. I arrived in Sault Ste. Marie just at dusk. My motel overlooked the St. Marys River; the famous Soo Locks allow shipping to pass from one lake to another. To the southeast, down the peninsula a ways, is Neebish Island, where Janet Lewis's father built the cabin where his family summered. There is now a handsome toll bridge across the St. Marys River; all night those great land ships, the trucks, pour into and out of Canada. The beauty here is the beauty of the waters, which split the immense forests and open the world to the sky again. If, in the west, one sees sky and land, here it's sky and water. I looked across the river at Canada for a long time—smoke from a factory on the north shore of Lake Michigan drifted across my view, blowing on across the prairie of water before dispersing.

The few residents of the Sault Ste. Marie whom I spoke to seemed jolly—it would probably be hard to convince most of them that any place could be more beautiful than the place they live.

The next day, after much studying of my road atlas, I decide to go west rather than south. Denied the Everglades, my southern options aren't good. The decision to go west, avoiding for the moment the vortex of Chicago, means a day mostly spent on the little roads so beloved of William Least Heat Moon and Annie Proulx. My route of choice is Michigan highway 28, a long road which allows one to slip between Lake Michigan and Lake Superior. For the purists who would like to see what rural America is like without the McDonald's or Pizza Huts which they so deplore, I recommend highway 28, which runs through a long stretch of the Hiawatha National Forest. The purists can eat, as I did, at a greasy spoon in Christmas, Michigan, where no one seemed to be feeling Christmassy just then.

Northern Michigan is the Maine of the midwest. The well-to-do of the midwest, at least the correctly well-to-do, had their summer places there; there is evidence along the lakeshores of the kind of expensive simplicity that prevails in the socially acceptable parts of Maine. Janet Lewis's father was a schoolteacher, Hemingway's a doctor; both had some sort of place up in Michigan. There is, however, another northern Michigan, just as there is another Maine: the Maine of the poor, as in *The Beans of Maine*, Carolyn Chute's vivid novel.° Driving west along the 28 I

°Or in the Michigan books of Jim Harrison, which I didn't read until after I made this journey.

saw plenty of Beans-of-Maine-type establishments. The northern forests are very dense along this road—now and then I would see where a family had not quite succeeded in hacking a homesite into them—sites not fully secured. The cars and household equipment had been sort of squeezed into the forest, only to have the forest shove a lot of it back onto the shoulders of the road. In some places Appalachian-type litter came almost to the edge of the pavement.

My drive through the Hiawatha National Forest called to mind not Longfellow but Henry Rowe Schoolcraft, the geologist, ethnographer, and administrator who was involved for so long with the Great Lakes tribes. He married into the Johnston family, the one Janet Lewis describes in *The Invasion,* and began to produce variegated and voluminous studies of Indian customs and Indian antiquities. Longfellow published *The Song of Hiawatha* (1855) one year before Schoolcraft came out with *The Myth of Hiawatha and Other Oral Legends,* but Longfellow knew the Hiawatha myth not only from Schoolcraft's massive studies but from the work of the anthropologist Lewis Morgan, whose critique of property and the primitive family so excited Marx and Engels.

As an ethnographer, Schoolcraft was one of the first convinced quantifiers—when he was compiling his massive, six-volume *Historical and Statistical Information Respecting the History, Conditions, and Prospects of the Indian Tribes of the United States of America* (1851–57), the questionnaire he sent to the various Indian agencies asked 378 questions. No sooner had I thought of Hiawatha than I passed the Gitche Gumee Campground.

It was misty and gray that day when I crossed Michigan, more misty and more gray once I began to follow the south shore of Lake Superior. A big tourist thrill along this road is a trip in a glass-bottomed boat to all the shipwrecks on the floor of the lake! It was not possible to enjoy the thrill quite so early in the season, though: the glass-bottomed boat might swiftly have become one of the shipwrecks.

It was so foggy around the lakeshore that I believed, for a time, that I was already in Wisconsin, when in fact I was still in Michigan, a state that extends around Lake Superior to within one hundred miles of Duluth, my initial starting point on this strange walkabout.

Once I did make it into Wisconsin, at Menominee, there proved to be no simple exit from the state. I could have gone on south, to Green Bay and Milwaukee, but by this time I was tired of the chancy sport of trying to pass logging trucks on narrow, fog-bound lakeside roads. Trucks there will always be, but I hoped to be done with the fog, so I cut across central Wisconsin, almost to Wausau, on highway 64, where I picked up I-39 and sped south; this meant bisecting Wisconsin from the northeast corner near Menominee to La Crosse, well to the southwest. As a result I had the opportunity to study Wisconsin more thoroughly than I had ever intended to. The middle part of the state, which I saw from highway 64, is a region of prosperous farms, much like those in Michigan or Minnesota.

Getting from northern Michigan to southern Wisconsin took the better part of a day—much of the day seemed dull, not because the country was uninteresting but because the fog and mist kept me from seeing more than a flash of it, here and there. Late in the afternoon I finally struck westbound 90–94, by which time I was tired and more or less without a game plan. To the west lay Minnesota and then South Dakota, a route that would soon put me back on the plains. My inclination was to go that way, rather than on south, though I was beginning to be a little concerned by the fact that what I mainly seemed to be doing on these drives was crisscrossing the plains.

The Mississippi at La Crosse, Wisconsin, though only about one hundred miles south of where I had first seen it at St. Paul, had already become a mighty river, cutting beneath impressive bluffs and spreading across a wide valley. I pulled off near the bluffs and watched the great river for a while, thinking about my tendency to be drawn back to the plains. Several long barges passed upriver as I watched—perhaps one or two of them were the same barges I had seen on the same river at Baton Rouge.

Then, as daylight began to fade, I turned west, and there were the plains again. I felt a little guilty for having returned to them so soon, but the guilt was foolish—I had no reason to avoid crisscrossing the plains. There was much historical precedent for doing just that. No sooner had the early explorers, adventurers, and traders crossed the plains than they turned and crossed them again. The Bent brothers crossed them tirelessly, as did Kit Carson, Jim Bridger, Custer, Cody, Charles Goodnight, Crazy Horse, Red Cloud, Quanah Parker. In more recent times writers as different as Mari Sandoz, Fred Manfred, Wright Morris, Ian Frazier, and Richard Manning had crossed and recrossed those plains. I surely have a right to do the same.

It's worth noting, also, that most of the great travel writers were compelled only by *one* landscape, almost always a harsh or difficult landscape: the deserts, the mountains, the rain forests, the poles. A few travel writers, such as Eric Newby or Dervla Murphy, are able to turn up almost anywhere and write charmingly about their journeys; but when most writers try that they only produce a kind of travel chat. With the possible exception of Burton, the great travel writers were specialists. Charles Doughty lived almost his whole life in a wet country but wrote his great book about the desert—the same deserts would later draw the best out of Wilfred Thesiger, St. John Philby, T. E. Lawrence, Gertrude Bell, and Freya Stark. Aurel Stein, Sven Hedin, Charles Marvin, Mildred Cable and Francesca French (the nuns of the Gobi), Curzon, and Ney Elias returned again and again to central Asia. Humboldt, Alfred Russell Wallace, and Henry Bates took their genius to the Amazon; while Mr. Darwin looked hard wherever he went. Certainly, when it came to those finches in the Galapagos, he looked every bit as hard as Picasso looked at Matisse.

It may be that America is too big a place to see whole. Perhaps I ought to be satisfied with the plains—and yet I'm not. Even as I start across them I'm thinking of the desert, where I plan to go in May.

Buffalo, Wyoming, where I-25 crosses I-90, is almost a thousand miles west of the bluff where I sit, watching the Mississippi. If I turn south on the 25 I could follow it much of the way home without losing the plains. West and south, I would cross almost twenty-five hundred miles of prairie country before pulling into Archer City.

Twenty-five hundred miles of plain is a lot to tack onto a trip that was mainly designed to allow me to see the country that Janet Lewis and Ernest Hemingway wrote about. I'm thinking that I might want to halve it, turn south on the 281 or the 83; but meanwhile it's dusk, I'm now in Minnesota, it's still raining, and I'm tired. A young Amish farmer, plowing with six fine draft horses, has just given up for the day and is following his team to the barn.

As I'm checking into a motel in Albert Lea, Minnesota, I see from a paper someone has left on the counter that two teenagers in Littleton, Colorado, have gone on a shooting and bombing spree at their high school, leaving (it was first supposed) as many as twenty-five dead. That number was later reduced to fifteen, but it still meant a lot of dead children; had the two young killers managed to ignite a twenty-pound propane bomb they had smuggled into the school's kitchen, they might have achieved numbers that were up there close to the 168 people killed in the Oklahoma City bombing.

But fifteen dead, fourteen of them children, in a school near Denver, is more than enough to make the plains not an inviting place to drive. I'm still haunted by a winter tragedy in Massachusetts—a little boy had gone out in a blizzard to find his dog; the dog came home but the little boy froze. I don't want to drive across the plains thinking about Littleton, a suburb that sits just where the plains end; nor do I want to read or listen to the endless probing for explanation that will fill the papers and the TV screens for the next week or two—until some fresher tragedy supplants the one in Colorado.

The authorities, stunned, are temporarily unable to believe that two teenagers, working alone, could smuggle half a dozen guns and something like thirty bombs—one of them a big bomb—into school without help, though it seems to me that that is just the kind of thing crafty teenagers *can* do, if they are of a mind to. The killing occurred on Hitler's 110th birthday; the young killers affected, it would seem, a kind of half-baked Nazism, which is as good an explanation for this tragedy as we are likely to get. Consider the conclusion of Susan Sontag's famous essay "Fascinating Fascism": "Now there is a master scenario available to everyone. The color is black, the material is leather, the seduction is beauty, the justification is honesty, the aim is ecstasy, the fantasy is death."

With the exception of the leather—the two teenagers just wore black trench coats—that's Littleton. The scenario was available to everyone, and these two rather nice-looking, rather sweet-seeming kids enacted it, to the ruin of themselves, their friends, their parents, their school, and perhaps, their community. They did not even seem to be particularly unhappy kids.

If I have to think of tragedy I'd rather think of the little boy in Massachusetts, who at least died heroically, trying to save his pet. The children of Littleton just died. The only more or less satisfactory theory of tragedy that we have—Aristotle's—presupposes mature characters with tragic flaws that compel the action. But the two murderous teenagers in Littleton were legitimately *immature* characters: just kids. We have no theory of tragedy that can say much about gratuitous murder, committed in the name of no real philosophy, no real politics: it's just murder, a Gidean act without meaning. As a parent my first thought is of the parents who, between daylight and dark of that day, lost their children and may never be whole or happy again.

I don't see that it helps much, or explains much either, to point the finger at our violent culture. All cultures harbor murderers, and massacre has not been exactly uncommon in Colorado: there's Sand Creek, the Meeker Massacre, Cripple Creek, Ludlow Station. If one is going to blame something it might as well be the species, or the gene.

When I awoke the next morning the plains were as sodden as

the grief-heavy citizens of Littleton. All was wet, all gray. I decided just to slip down to Omaha and fly home. For about an hour of my drive, from Sioux City down through western Iowa to Omaha, Nebraska, I am running parallel to the Missouri River. In Sioux City, I believe, it is possible to go over to the stockyards and order a seventy-two-ounce steak; if consumed entirely, within one hour, the steak is free to the patron.

On a better day, with the plains less sodden and the news less troubling, I would greatly have enjoyed this stretch of the river, along whose very banks the first keelboaters, under Manuel Lisa or Major Ashley, had laboriously pulled their boats up the river, hoping to float down it again with those boats filled with furs—a dream that was seldom realized. There were furs aplenty up this great river, but getting them past the Mandans and the Sioux proved not to be easy.

Omaha, on the west bank of the Missouri, is a plain little town, so unassuming that it still calls its airport an airfield. I am confused by this, thinking it is only an airfield for private planes. Then I saw a large plane take off and realized my mistake. Since the airfield is right by the Missouri River there is no need for me to penetrate far into Omaha's bumpy hills. It seems strange that there should be a man here, Warren Buffett, who is worth some $50 *billion;* it's difficult to associate plain, unornate Omaha with such riches. Mr. Buffett is the stylistic opposite of the sultan of Brunei, another very wealthy man.

M A Y

San Diego to Tucson to Archer City
on the 8, the 10, the 20

IT MAY BE that the availability of speedy travel has mainly worked to make the human animal—or at least the American animal—more impatient. This opinion can be confirmed by walking through any large airport on a bad travel day. Getting places quick is a habit so rarely thwarted now that when it *is* thwarted the shocked travelers almost immediately go to pieces.

I decided, on a very soupy Sunday, to fly out to San Diego and drive home. I had had enough of the beautiful but murderous plains—what I wanted was the desert, more than one thousand miles of which lay between southern California and central Texas.

If there's one place where speedy travel is almost always available, it's the Dallas–Fort Worth International Airport—DFW for short. The airport is situated on a broad plain, where the visibility is almost always good; it receives and dispatches more than two thousand flights on a normal day. Only once in my life have I been delayed more than an hour or so there, and that delay occurred when a hurricane was making its way up the eastern seaboard.

There was no hurricane in progress when I arrived at the airport on this warm, rainy spring Sunday, but the east coast was experiencing weather severe enough to snarl air traffic nationwide. I knew this, and did not expect an on-time departure; I came prepared to wait awhile. Even so I was surprised by the size of the crowd that had jammed itself into the airport by the time I arrived. Something like 150,000 people were temporarily stuck, every one of whom, it seemed, had been expecting—indeed, had been counting on—an on-time departure. Though most of these Sunday travelers had only been waiting about an hour, it was clear from their desperate behavior that they had already assigned themselves the status of stateless refugees. This was made more evident by the fact that all the TVs in the American Airlines lounges were tuned to CNN, which persisted in displaying the miseries of the *real* refugees who were in the process of being shoved out of Kosovo.

Though all this rain delay really meant was that a lot of people would be a couple of hours late getting back to Newark or San Jose, the people being asked to accept this brief delay looked, if anything, more bleary, more resigned, more despairing than those thousands on the Albanian border whose lives were being destroyed. Thanks to CNN it was impossible not to compare the two groups: those Americans who were certainly going to get home, though a little late, and the Kosovars who no longer had homes. Both groups sat amid heaps of possessions—in the Americans' case, whatever they had taken on this particular trip, or bought during it; in the Kosovars' case, all they now owned in the world—but on the whole, the Kosovars looked less defeated than the Americans. They looked alert and resilient, prepared, since they had no choice, to accept a state of war; but the Americans, who also had no choice, were not at all prepared to accept a little travel delay.

To be fair, not all the stranded Americans looked resigned and defeated; a few refused to concede a thing to circumstance and kept harrying the airlines, determined to find a way out of Dallas. I coped with this delay by giving the Sunday *Times* an exceptionally thorough reading—while I was about it, two forceful ladies from Newark returned to the check-in counter three times, determined, by sheer will, to force a passage through the clouds and get *somewhere*. What about Cleveland? they asked. What about Baltimore? When turned down for both locations they went away shaking their heads, unable to believe that a great airline such as American could not even deliver them to Cleveland.

When I finished my *Times* I strolled about for a bit, amid the masses of the stranded, many of whom were so dulled out by now that they had ceased making any effort to restrain their children, who were, to put it politely, running riot. I wanted to see if I was still good at an old game, one I had sometimes played when I flew several times a week: I would try to guess the destination of a given flight just by the look of the people who were about to board it. World-weary cynicism meant Baltimore, hyperactive cynicism Miami or Las Vegas. Trench coats meant Washington, D.C. The gray people, whose cheeks no sunlight seemed ever to have touched, were usually bound for Cleveland or Chicago. A certain stridency of voice meant Newark, or maybe La Guardia . . . and so on.

After playing the game happily for a few minutes I concluded that the hardest passenger groups to identify definitively were those bound for the Pacific northwest. People bound for Portland or Seattle do not seem to have acquired much regional coloration. Perhaps, at most, they betray a slight tendency to look collegiate. I once bumped into their First Citizen, Bill Gates, in a hotel in Tucson and even *he* looked collegiate, for all the many billions that he commands. The Pacific northwest remains determinedly downscale; people headed that way wear, almost always, a touch of gloom, but modest, civilized, Scandinavian-type gloom.

While wandering past the airport bookshop I noticed on prominent display Michael Korda's new book, *Another Life.* There it was, its shelf position even better than that of John Grisham's latest novel. Michael Korda has been my editor for more than thirty years—only the week before, he had sent me the page proofs of the very book on display. That Michael has written a charming book is not amazing, but that his book traveled from page proofs to an airport bookshop in one week *is* amazing. Yet there it was, a book.

My flight to San Diego took off about two hours late but still plopped me down on the west coast with plenty of time to look around. I first drove down interstate 5 almost to Tijuana, in order to have a look at the border at one of the places where the pressure of immigration is greatest. Once there, I felt something of the same unease I feel on the Texas or Arizona border. Though not in Mexico, I *was* in the pressure zone, the area the poor of the south have to cross if they are to make it to the American Eldo-rado. I then went as far west as it is possible to go by car, to the beach at Border Field State Park, where I looked at the ocean for an hour—gray water under a gray sky.

My drives across the American land had taken me far enough that I had begun to feel a vague urge to try a different mode of travel. For the past month or so I had been reading the leisurely, tolerant travel books of the English zoologist F. D. Ommanney, a man who knows a lot about fish, and a lot, also, about the world's oceans and the people who live beside them—particularly the island peoples of the Indian Ocean and the South Pacific. F. D. Ommanney was a fish finder, a man who, in the years after World War II, puttered around in remote oceans attempting to estimate whether a given stretch of ocean contained fish enough to make commercial fishing profitable. I think, though, that what he cared about was the sea, not the fishing. In books such as *A Draught of Fishes, The Shoals of Capricorn, Eastern Windows,* and *South Latitude,* he describes his journeys through the seas and islands so appealingly that a landlocked person such as myself begins to feel that he has really been missing something: that is, the world's oceans, along whose trade routes—invisible highways—the great ships proceed.

The appeal of F. D. Ommanney's books—fairly popular in the 1950s but mostly forgotten now—is their intimacy with the sea and its ways, and also with the ways of people whose lives are bound to the sea. Conrad and Melville wrote powerfully of the oceans, but their works don't exactly bring one into an intimacy with the world of the waters. In Conrad and also in Melville the sea is too powerful, too often the environment of crisis, to be merely appealing. Though these great writers see the ocean's beauty they rarely allow the reader to be unaware that this beauty comes with a threat, moral or physical or both.

Ommanney is not a novelist—he is just a man with a deep interest in the natural world, particularly with the world of the ocean; through many travels he preserves a fond curiosity about the lives of peoples of the islands, people who can scarcely imagine a life apart from the sea.

Reading Conrad or Melville has never made me want to get on a boat and venture out on the prairies of water, but that's the impulse reading Ommanney produces. I don't want a fancy boat, either—not one of the big liners that would make Nelson Algren envious if he happened to see William Styron getting on it with his wife Rose. I would prefer a more workaday boat, of the sort Evelyn Waugh seemed to be able to book when one press lord or another sent him off to Africa to report on a war. There were a couple of such ships just visible from where I was strolling—big commercial boats, far out on the horizon. For a moment I felt a keen desire to be on one of them, bound for the mystic Orient.

The sun was hidden that day—sunset was only a pink glow behind the clouds; the outlines of the great ships melted into the grayness and were gone. I got back in my car and drove north to the popular beaches, from which there was, at this hour, a steady exodus of sandy people and tired children. Some people, though, weren't leaving; they sat with their towels wrapped around them, watching the ocean as darkness came.

I have been coming to San Diego for more than thirty years, but always as a book scout. If I saw the ocean at all it was only a glimpse or two, as I sped down the highway. Almost a decade earlier, depressed after heart surgery, I was encouraged by my psychiatrist to go live by the sea. It might soothe me, she thought, and she was right. I moved to Santa Monica, less than two blocks from the ocean—though I was up on the cliffs rather than down on the beach. Nonetheless I woke up every morning smelling the spray and walked every day by the sea. It did soothe me, and it also came to interest me. I began to read sea books, but found that most sea travelers were as obsessive as the great desert travelers, the great polar travelers, or the great mountaineers. The seas became merely fields of water upon which their personalities operated. Ommanney, by contrast, was a happy discovery; he is always more interested in describing what he sees than he is in examining himself. The photographs in his books are casual and amateurish but the observations he makes are precise and unusually generous. Though occasionally exasperated, Ommanney likes island people and does full justice to their attitudes and virtues.

In the morning, when I set out to drive east on interstate 8, it is so foggy that I can barely find my way off Harbor Island and onto the freeway. Fog or no fog, traffic is already aboil. Just as I curve onto the 8 a truck passes me bearing a large sign to some new location. The sign merely says: SELL YOUR BABIES, with a phone number. There are no embellishments on the sign at all— no suggestion that selling your baby will provide it with a stable home, or anything else. SELL YOUR BABIES is what it says, and all it says.

If that's not a rude enough start to the day, I then glimpse, glimmering through the fog, some powerful neon, advertising a casino with an outlet mall attached—an idea, I suppose, whose time was bound to come.

Those two signs, plus lots of taillights, are the only things I see for more than an hour, as I proceed cautiously east through the mountains that lie near the coast. I knew from previous drives that the scenery is splendid around Alpine and Pine Valley, but I am not allowed even a glimpse of it today; the fog is so thick that I consider stopping, a thought that doesn't seem to have occurred to any of my fellow travelers, who keep whizzing past me as if they were enjoying normal visibility. Perhaps it *was* normal visibility for those heights at that time of day. I hew to the right-hand lane and look sharp, but even so almost rear-end a truck that is struggling to pull a long grade.

The fog held for about sixty miles, until I began a rapid descent into the desert, near Campo, California, beyond which I soon exchanged fog for blowing sand. Quirky winds began to jerk me this way and that, as I came out of the mountains. A large dust devil was just crossing the highway, whirling roadside debris high in the air—the dust devil was a tiny cousin of the great family of tornadoes that, a day later, destroyed a big swatch of suburban Oklahoma City.

When I drove the 8 twenty years ago, shortly after it opened, much of the area between Gila Bend, in Arizona, and El Centro, California, looked like Arabia, dunes stretching everywhere. There are still some impressive dunes in the Imperial Sand Dunes Recreation Area, between El Centro and Yuma, but big agriculture has dribbled Colorado River water through the rest of this desert and converted it into huge farms. The fields—more than two hundred miles of them—are lush, with only the sandy ditches to remind the traveler that he is in a desert.

I started my day at sea level in San Diego; rose, in the mountains, almost to five thousand feet; and descend back to sea level before I reach El Centro. The mysterious, lovely Salton Sea, which is actually below sea level, lies just to the north. There are many old desert couples who return year after year to the Salton Sea; they park beside it in their ancient campers, wade in it, watch the pelicans.

The wind whips long skeins of sand off the Imperial dunes as I pass through them—ahead, the lower quadrant of the sky is beige. The northern beaches of the Gulf of California are not far south, which is why, at times, the desert feels so humid. Pickers are in the fields, but so far away that I can't see what they are picking. The Colorado River at Yuma is very green.

I am not one to do much breast-beating about environmental change or even environmental destruction. The human species is clearly a species that uses up its habitats, recklessly and greedily, just as big agriculture is now using up this desert and the finite waters that can make it flower. What feels wrong about it as one crosses from El Centro to Yuma is the scale. The tribal peoples of the desert, the Piman peoples, have always found ways to water the desert, else they couldn't have survived in it. As desert agriculturalists they worked with great skill, but on a modest, intimate scale, whereas today the scale of agribusiness in this desert is immense; the fields between Yuma and Gila Bend are as large as any in the midwest. Driving through them produces a nagging sense of dislocation. What is all this cotton doing here? Isn't it in the wrong place? I remember feeling something of the same dislocation—of sadness, even—when the Texas panhandle, the great range where my father and my uncles used to go dashing in their days as cowboys, began to be plowed, irrigated, and turned into cotton fields. I wanted those prairies to be left alone, though for no better reason than that my family had been formed on them. But we're a grasping species; nothing of value is likely to be left alone. If the land can be changed so as to be made to yield more money, it will be; though eventually, when it no longer yields money, it may slowly go back to being what it once was. The water may play out, both in the desert and on the plains; the one may go back to being sand dunes, the other to being prairie. All human intentions are, in their way, short term—in the context of today, this is a consoling thought. In the great media markets of Hollywood and New York the trend is toward consolidation; every year there seem to be fewer and fewer players controlling ever larger and more complex entities. Soon the whole world of entertainment may belong to Michael Eisner, Rupert Murdoch, Ted Turner, and one or two others. What is forgotten is that all these major players are aging men; what they have so competitively gathered together may, in only a few years, scatter again, break back into fragments. The entities they have created are for a day, not forever.

I'm happy to take the long view, which allows me to imagine a day when these misplaced cotton fields will be shifting sands again. The fact is, I'm tired of the fields; so, at Gila Bend I decide to make a detour. Rather than drive two more hours through the fields I drop south on highway 85 toward the border, which is about one hundred miles away. This route takes me through the Barry M. Goldwater Air Force Range, where, most days, young fighter pilots can be trained in the use of smart bombs and other reasoning weapons.

It's not happening today, though, perhaps because the air force has received reports that there are too many illegals from

Mexico straggling through this huge desert range in hopes of reaching an American city, where they might find work.

These reports are accurate. I have hardly gone five miles south, into the desert, when I see one of the tiny Border Patrol helicopters at rest just off the road. Also at rest are ten Mexican teenagers, sitting on an old railroad tie—they have been run to earth by the helicopter and look tired and dejected. They are nearly one hundred miles inside the Promised Land and, with a little luck, might have made it the few more miles to Gila Bend, where some farmer might have hired them to pick something. Just as I pass them, the white deportation bus comes up the road, in a hurry to take the kids back to the country they have just walked out of.

I see this sight twice more, in the 150 miles I have to cover before I reach Tucson. East of Ajo I see two teenagers who have just managed to straggle out to the highway. They appear to be near collapse. One sinks down even as I approach; he is at the end of his endurance. The car in front of me pulls over, and I do too, but before either of us can come to a full stop a Border Patrol vehicle swoops over a little rise and bears down on the boys. Two more of the small Border Patrol helicopters, looking like small dragonflies, are hunting in the area, one to the east and another to the south. Against such technology, the kids and the families who make their way north would seem to have little chance—but thousands keep coming, and some get through. Even the Border Patrol can't be everywhere. All along two thousand miles of border, from the Pacific Ocean to the Gulf of Mexico, such scenes occur every day; in California and Arizona the Border Patrol seems mainly to be trying to turn back immigration, whereas in

Texas the same force is bent on turning back drugs as well. Driving up to a Border Patrol kiosk at night in west Texas usually produces a small tic of anxiety, even in legal citizens. The drug dogs, loosely restrained, are almost always German shepherds; however mild and courteous the Border Patrolman may turn out to be, the image that registers in the mind, as you drive slowly toward the uniformed policeman with the big dog, is an image from the iconography of Nazism. The association might be unfair, but it's also unavoidable. The struggle of the poor brown people of the south and the more affluent northerners who seek to retard their entry is unrelenting and, as I said earlier, corrosive. It poisons the whole border, makes it not a pleasant place to be.

Just before I come to Ajo I notice an odd geological development, an area of tiny buttes, most of them only twenty or thirty feet high, which look as if they could be models for the great buttes of Monument Valley. It's as if Monument Valley had been miniaturized and set down in the Sonoran Desert, for those who haven't time to visit the big valley to the north.

As I come into Ajo, an old mining town and a grim one, I see a sign that says, "Leash Law Enforced." Leash law, in Ajo, Arizona? I wonder if the law is to protect the dogs, or the citizenry. Ajo is a tough place—with several hundred well-established coyotes within a mile of the town, how long could an unleashed dog expect to last?

A few minutes later I pass through Why, one of the most appropriately named communities in North America. Most people, finding themselves in Why, will be likely to question the fact that they are there. Why is hot, dry, dusty, and without amenities of any sort. Some of the trailer houses scattered around the outskirts have been there so long that they look like volcanic encrustations of some kind.

There is, however, the brand-new Golden Ha-san Casino on the road to Tucson—it's tiny but could be considered an amenity, I suppose. The casino is located right on the edge of the Tohono O'odham Reservation. What makes it especially conspicuous in Why is that it's clean as a pin; nothing else in Why is that clean, not even the pins.

The road from Why to Tucson is almost as straight as an arrow—it has plenty of dips but few curves. It's a lonely road whose emptiness and straightness often tempt drivers to fatally high speeds. In the 120-mile stretch between Why and Tucson I count thirty-two roadside crosses, places where someone has died. In several places there are two crosses together; in one place, three crosses. Several of the death sites have small, well-tended shrines, ringed by pretty stones. One had a Madonna inside.

The frequency of death along this excellent road discourages one, though. Cars pass me as if I'm not moving, although I'm driving eighty. When still about seventy-five miles from Tucson, I can see Baboquivari Peak, 7,730 feet high, the sacred mountain of the Piman peoples, the desert-dwelling Indians now called, by their preference, the Tohono O'odham. Near Baboquivari is Kitt Peak, home of a famous observatory. For the next forty-five minutes I have both peaks in view.

Just as I come to the outermost intersection on the west side of Tucson, I see that the highway has just made a strong claim on yet more victims. Four cars, none of them traveling slow, have collided in a kind of wild frenzy. There's a helicopter there, several ambulances, and a small squadron of police cars. It's hard to tell, as I inch past the wreckage, who's dead, who's alive, or what happened. An element of contest is not uncommon in western driving—that's one reason it's so dangerous. It may be that all four drivers were trying to beat one another through the intersection, the result being that no one made it.

Four schoolchildren, just out of school, have other things to think about, though. They walk with their book bags past the cops and the milling crowd, mildly annoyed that their school bus has been forced to stop far down the road. The car wreck doesn't interest them.

One of the glories of Tucson is its early morning light. Even the first tentative graying of the sky has a bright quality, and in early May it's light well before 5 A.M. When the first sunlight spills over the mountains it brings an hour of quiet, cool clarity. From the foothills of the Santa Catalina Mountains one can see across the whole wide valley and on far to the south, almost to Mexico. As the heat increases, the light changes, becomes heavier. People in Tucson come to appreciate the clarity of that early light, as wine drinkers appreciate a special bouquet. Though never quite my home, Tucson has been a favored stopping place for years, in part because of the quality of its light. When I was there recovering from heart surgery, the way the light seemed to flow over the mountains—a river of bright air—brought a lift that I came to depend on. During the hot months the sky blurs during the heat of the day—clarity only returns in the evening. The light at sunset has deeper tones than the light of morning, but I still prefer the sunrise. Starting one's day with brightness falling from the air—as in Thomas Nashe's poem—is tonic to a sun lover.

I lingered in Tucson for three days, enjoying the spring light and pondering what route to take as I proceed on eastward, toward home. I had meant to go northeast through the Salt River Canyon and on up to I-40 at Holbrook, where I thought I might peep into the famous Meteor Crater. I had been looking forward to going up on the Mogollon Rim and crossing the great northern Arizona plateau toward the lands of the Pueblo people, starting at Zuni and going east.

Edmund Wilson, whose journals I've just been reading, was particularly fascinated with Zuni and wrote a long, vivid account of the Zuni ceremonials. It's in *Red, Black, Blond, and Olive.* For one so down east, Wilson spent a surprising amount of time in the west—in California, Nevada, New Mexico. He was a great describer—after all, he even described *Finnegans Wake.* His travel pieces make excellent reading today.

What thwarted my plan to climb the Mogollon Rim was my car. The luck of the rental car draw had stuck me with a car that wouldn't readily go uphill; efforts to exchange it in Tucson proved fruitless. I didn't want to get down in the Salt River Canyon with a car I might have to push up the other side. Attempting to twist my way up the rim with a car so underpowered that it couldn't zip around some faltering Winnebago, should opportunity offer, did not appeal. Few opportunities for easy passing *do* offer in that part of the country, which can take a long time to cross if one is stuck behind a line of RVs and Winnebagos, and in this season there would surely be herds of them to somehow get around.

The conclusion I drew was that the Salt River Canyon, the Meteor Crater, the Mogollon Rim, and the many pueblos would all still be there later in the year, when I might have a better car.

So I pulled back onto the 10 and followed it east, as I have many times in my life. There is a certain relaxation that comes when one drives an often-traveled road, especially if it's a road that goes through beautiful country, as the 10 does between Tucson and El Paso. Apart from the bump of Texas Canyon, a short, boulder-strewn massif just east of Wilcox, Arizona, the 10 is a flat road through beautiful desert country. There are few farms along it—the Colorado River is now out of reach; the mountains are always in sight but never in the way. Near Wilcox there's a famous tourist stop advertising THE THING—in fact an Anasazi mummy. West of Lordsburg, just as one is entering New Mexico, there's a flat sandy basin that can present problems for trucks if the wind is really blowing. It was really blowing the day I drove home—two eighteen-wheelers had been flopped over by it; they lay on their sides, like exhausted beasts, being studied by a number of wrecker drivers, who were trying to figure out how to get them up and going again.

A few years ago I crossed between Lordsburg and Deming on a day when the wind was blowing so ferociously from the north that it produced a tumbleweed stampede of unique dimensions—unique, at least, in my experience. Tens of thousands of tumbleweeds had broken loose in the sage country north of the 10; the tumbleweeds were sweeping down on the highway like an alien horde of attack plants. Some were as big as Volkswagens—indeed, I saw a giant tumbleweed hit a Volkswagen and engage with it so tightly that the driver of the Volkswagen had to stop and pull it off. There were so many tumbleweeds crossing the road that there was no dodging them all—and thousands more were coming, crossing the desert in great bounds which left them airborne for twenty or thirty yards. I myself soon hit a large one—I still had part of it stuck to my grill when I reached Archer City, eight hundred miles later. The wind was high and the tumble-weeds ready to rumble, or tumble, that day.

Nothing as dramatic as a tumbleweed stampede happened on this drive. There was just enough dust in the air to blur the land-scape ahead. Usually, when approaching Las Cruces, New Mex-ico, I can see the jagged, mini-Tetons-like range of the Organ Mountains from at least forty miles away. Today I couldn't see them clearly until I was coming down the long slope into the Rio Grande valley, only a few miles west of Las Cruces.

The sixty miles from Las Cruces through to the eastern edge of El Paso are the only stretch between Tucson, Arizona, and Pecos, Texas, that I don't enjoy driving. Between Las Cruces and El Paso there are feedlots, factories, suburbs blocking my view of the Rio Grande. After a long stretch of driving across prairie and desert, where the traffic is sparse and one can just look and think, the sudden demands on one's attention made by the traffic through any large city will feel like an unwelcome imposition. From being almost alone, one is suddenly amid five or six lanes filled with automotive killer bees. The driver may know it's com-ing, but even so, it's a shock.

A singularity of the 10, as it slices through downtown El Paso, is that for a mile or two it bumps one right up against the Third World. The Rio Grande runs just beside the 10—for a mile or two one can look across into Juárez, only yards away. Kids will be play-ing on the dusty hillsides, and most days, a few people will be casually wading the river, their shoes slung over their shoulders. A riverfront several miles long through the heart of the two cities presents the Border Patrol with a challenge it can't meet. Crossing the river is just a daily fact for many Mexicans, and many Americans too—it would take a large army to stop it. The First World traveler, cruising in luxury on the 10, can look out the window and see, briefly but very clearly, how the other half lives— only, in this case, it's a lot more than half. Juárez has grown much larger than El Paso—at night its lights stretch so far into Chihuahua that driving past it is like driving past L.A.

Van Horn, Texas, is the first town inside the Central Time Zone, a zone which stretches an immense distance, across America, over to eastern Tennessee. Van Horn survives mainly as an oasis for truckers. As I'm coming down the long slope into town, a little after dark, I count twenty-six sets of taillights ahead of me, all on trucks and all distinct in the clear desert night. I'm going to stop and sleep but most of the truckers will keep on trucking. By the time I'm on the road again the next morning, most of the truckers will be in Dallas.

There are towns in the west where it's not even wise to roll down one's windows—if you avoid getting murdered you might still breathe in some deadly desert germ. For a long while, I considered Van Horn to be one of these towns. Yet two years ago, I spent some weeks in Van Horn, overseeing the production of a miniseries, and survived, an achievement made easier by the cooking of an admirable woman named Rosa, who has put a son through graduate school at Boston University by cooking what is probably the best food between El Paso and Fort Worth. Rosa cooks in a small Mexican-food restaurant just by I-10. Noble human achievement is fortunately not confined to the pretty places of the earth. Van Horn is one of the grittiest towns on the whole length of the 10; few places are less like Boston—and yet, still, the spreading sunrise makes the great empty desert look beautiful for an hour or two, after it rises.

Though I'm now well into Texas, I'm still five hundred miles from home, a fact that prompts me to study my road atlas for a while, seeking a road to the north that will allow me to skirt the ugly, oil-pocked Permian Basin, which I will strike about one hundred miles east. Oil in vast quantities has been sucked out of the Permian Basin since World War II—some notable fortunes have been made in it. But the detritus left by more than fifty years of oil drilling has turned an indifferent transitional region—not quite desert, not quite plain—into a kind of oil hell. Baku is probably uglier, but Baku doesn't lie along the 10.

The roads north from Van Horn or even Pecos are not promising; they would throw me back a little west, toward Carlsbad, New Mexico. North of Carlsbad, in the tiny agricultural community of Lake Arthur, New Mexico, some twenty years ago, the face of Jesus appeared on a tortilla a poor Hispanic woman was cooking. Her life had not been easy, but became easier once the Savior came and blessed it. Her modest house, in this very poor town, became in the eighties a major shrine, visited by thousands. The Holy Tortilla, as it came to be called, was carefully preserved; when the woman had to leave to run errands she would tack a note on the door which merely said: "Come In And See Him." I was there once—I went in and saw Him. I suppose the Holy Tortilla is the American equivalent of the Shroud of Turin. What was impressive was the care the woman took to see that all who made the lengthy journey to Lake Arthur would be able to see Him whether she was home or not. I imagine He is still there, and that the note is still on the door.

My studies in my road atlas proved unfruitful. Apart from the blindingly white Salt Lakes north of Van Horn, there was really nothing to tempt me in that direction. It was far too windy to contemplate going north to Lubbock—once I struck the vast stretches of plowed land south of there I would merely be driving through a wall of sand. So I went on east along the 10 until it angled south toward San Antonio, Houston, and the Petroleum Coast. At that point I joined the 20 at its western beginning and/or end. The 20 goes on and on, past Fort Worth, Dallas, Atlanta, until it finally merges with the 95 in South Carolina.

The 20 doesn't offer beauty or interest, but it does provide the cardinal virtue of all interstates: speed. I zip along, secure in my belief that the communities of Toyah, Pecos, Barstow, Pyote, Wickett, Monahans, Penwell, Odessa, Midland, Stanton, Big Spring, Coahoma, Westbrook, Colorado City, Loraine, Roscoe, Sweetwater, Trent, and Abilene require no inspection. They rarely do require inspection, though the famous rattlesnake roundup photographed so vividly by Richard Avedon in *In the American West* occurs annually in Sweetwater. Many thousands of reptiles, whose skins will become belts or boots, are collected there each year. This harrying of the rattlesnake—as many commentators have noted—brings in plenty of tourists but is not environmentally wise. Thanks to the big snake hunts I have not seen a rattlesnake near my ranch house in more than ten years, but I see plenty of the rats and mice that the rattlesnakes would be eating, if only they were still there.

JUNE

L.A. to Albuquerque on the 10, the 15, and the 40

WHEN FLYING INTO L.A. I always try to glance down as the plane crosses the San Diego Freeway, now usually called the 405. If it's not too smoggy I can tell at a glance if the traffic to the north is flowing: even the speed of syrup is good enough. From that hasty look I usually decide what route to attempt, once I get my rental car and head north toward Beverly Hills. Do I trust myself to the ever chancy 405, or just plod up Sepulveda, a route that's dull but safe?

Where the freeways of Los Angeles are concerned, the wise man takes nothing for granted. It takes a while to land at LAX, debark, secure a rental car—twenty minutes at least. In that length of time a bobble or small accident almost anywhere in the system—on the 10, on the Harbor Freeway, on the 405 itself—can bring all traffic within ten miles to a sullen halt. By the time I get a car and come whipping off Century Boulevard, or Manchester or La Tijera, all bets may be off, the 405 so jammed that it is reluctant to accept even one more vehicle.

For some years I kept an apartment in Sherman Oaks and arranged, on most occasions, to fly into the once-quaint (now conventional) Hollywood-Burbank airport, avoiding the great but frequently conflicted San Diego Freeway altogether. It was a nice apartment. During the years when my screen work was done mostly for the Valley studios—Universal, Warners, Columbia—having an apartment a few minutes up Magnolia Avenue was quite convenient.

But people who allow their lives to be determined by convenience are not the sort of people who flourish in L.A. The city is a lot of things, but it's never convenient—except, perhaps, for its eccentrics, one of whom was a little old man who used to drive up and down Santa Monica Boulevard in a car shaped like a yellow shoe. He was a cobbler—it probably felt normal to be driving to work in his shoe car; but for the rest of us, it was a sight to see.

Long before I gave up my apartment in Sherman Oaks I ceased to use the apartment, or even to go to the Valley. I resumed my old habits, which meant flying into LAX, struggling up the 405, staying at the Beverly Wilshire. I kept the apartment mainly because I liked my landlady; but also, probably, in the back of my mind was the thought that an apartment in the Valley was an inexpensive hedge against the day when I might be an old for-hire screenwriter—Gore Vidal's Wise Hack—working on unimportant movies at Warners or Universal.

Los Angeles is not a simple town. David Rieff, labeling again, calls it the capital of the Third World. It's true that it contains such a mixture of peoples now that coming into it is like coming into a country, rather than a city; but my own experience with it has not been with its Third Worlders. My experience, to the extent that I could arrange it, has been with the Ultras, the great stars, directors, beauties: people who are not just First World but a stratum above the First World. The Ultras are the people whose job it is to create desire on a global scale—desire in all those everywhere whom the movies touch. It's a compound desire, too: for sex, for money, for glamour, beauty, style. L.A. may have oil money, aerospace money, computer money in billions, but that's not why people go there. They go to L.A. for stars—the hope of being a star or, at least, of seeing a star still animates the multitudes in L.A. That's the prospect that keeps alive a state of desire.

I first came to L.A. as a book scout in 1959. Two years later, my first novel, *Horseman, Pass By,* having been published and filmed (as *Hud*), I made my first trip there as a screenwriter, at the behest of the late Alan Pakula, then a producer. Since the early sixties I've been in L.A. often, sometimes just as a book scout, other times mainly as a screenwriter. In my hours off from story conferences (now called meetings) I've bought lots of exciting books, courted one or two beauties not very successfully, eaten, walked, and stayed at such varied hostelries as the Biltmore, the Ambassador, the Hollywood Roosevelt, the Chateau Marmont, the Beverly Wilshire, the little-known Beverly Rodeo, and the Sportsmen's Lodge, on Ventura Boulevard in the Valley.

What made this trip different was that, for the first time in forty years of visits, I had come to Hollywood looking for work:

always previously, as reported in my little book *Film Flam,* work came looking for me. I'm not quite sure why it stopped looking for me—perhaps it's merely the curve of the years. Two years ago, while living in Santa Monica doing postproduction on a miniseries of one of the *Lonesome Dove* books, I noticed a curious thing. The studio executives I would go and talk to about one project or another were seldom even half my age. Now they were only a little more than a third my age. I was in my sixties, they were in their twenties. Some of them seemed puzzled that an older person would still be writing screenplays. If I happened to mention, by way of illustration, a movie made as long ago as the 1950s—twenty years before any of them were born—they looked blank and, in some cases, a little disdainful. I might as well have been talking about the Dead Sea scrolls. There is always a listener (the executive) and a note taker at these meetings. If I mentioned *Touch of Evil* or *Roman Holiday* the note taker would dutifully take a note.

I don't know why this age gap surprised me. Hollywood, as I said, has always been about beauty and desire, neither of which is entirely comfortable with age. Garbo was not wrong to retire.

THE FREEWAYS of Los Angeles have been, to some extent, my Ganges; for forty years I've been making my way slowly up or down them, stalled about as often as the Newbys were on the holy river. But this time the 405 was flowing smoothly; it carried me north and deposited me like a grain of silt on Santa Monica Boulevard. Lots of other grains of silt were deposited with me, of course. I didn't have to bother about the freeway system again until I was ready to leave. My writing partner, Diana Ossana, came in from Tucson, we stayed at the Beverly Wilshire, and we never went east of it, even though we conducted eight meetings. It just happened that all eight lay to the west, from Century Park East to Fourth Street in Santa Monica.

In the old days, as I've said elsewhere, the most difficult part of screenwriting was just getting onto the studio lots. The security guards, past whom you had to penetrate, were all possessed of an intense, almost Balkan suspiciousness; gaining entry to a studio was like trying to get into a Communist bloc country, when there was a Communist bloc. Papers were scrutinized for imperfections; phone calls were made to commissars deep in the lot.

Getting into studios is still not easy, but nowadays, in the era of independent production, it's rarely necessary. The town is dotted with production companies, small and large, tiny nation-states that have broken away from the control of the old politburos. Each presents its own difficulties, but one difficulty they all tend to have in common is parking. People talk about Internet billionaires, but what about Valet Park billionaires? Hollywooders may be willing to spend their lives on the freeways or beside the freeways, but one thing they are increasingly reluctant—indeed, unable—to do is park their own cars. In Santa Monica, West Hollywood, Beverly Hills, Westwood, even laundries have Valet Park. Grocery stores have Valet Park. Beneath the great office towers of Century City armies of small brown men wait in the depths of the earth, stationed there to whisk your car away the moment you step out of it. A few minutes later, when you're finished with your meeting, they'll whisk it back, at a cost of only about fifteen bucks.

Valet Park is a phenomenon that's come on fast. Ten years ago even the Ultras would occasionally have to park their own cars; now such a thing is unthinkable. No high-end restaurant could survive a week without Valet Park. The eccentric patron who now and then insists on parking his or her own car is looked upon as a dangerous radical, a threat to the new economic order.

I don't mind Valet Park, particularly—in fact I'm thinking of building a novel around it. I see it as the newest *real* advance in California living; L.A., as usual, is on the cutting edge.

What I now hate are the great garages—these garages have replaced security guards as the main threat to the serene life of the screenwriter. In Century City particularly these garages are Dantesque, leading one inexorably down, level by level, into a hell of whiteness, in which all the parking spaces seem to be reserved, either for the legions of the handicapped or for the even more prolific legions of executives, somewhere in the tower above you, for whom a secure parking space is simply one of the perks of the job. In Century City these garages are apt to suck you in off one street and spit you out onto quite another. I've always rather prided myself on a plainsman's sense of direction, which usually enables me to tell east from west and north from south. Century City stripped me of this sense completely. In fact, Century City itself, a world of great towers stuck there between Beverly Hills and Brentwood, is essentially adirectional, taking its character from its main street, the Avenue of the Stars. In Hollywood, stars have levels, not directions; they are, by definition, visible from all directions.

But screenwriters, no. Those vast parking garages, in which thousands of places are either filled or reserved, should only remind the reflective screenwriter of the industry's long ambivalence about them: they may have invitations, but they don't have parking places—not ones, at least, that they can call their own.

The offices of today's production companies, whether the company is aiming to produce feature films or fodder for TV, are remarkable for the high impersonality of their design and decor. Though there are frequently humans in them, the offices show no signs of human use; in the several that we went to I was always by far the untidiest person in the place, and this on a day when I had more or less dressed up. Faced with so much scrubbedness, I felt rather like Pigpen, in the Charlie Brown cartoons. Not only was I a lot older than anyone we saw, I was also infinitely less groomed. I am, as my partner would admit, essentially ungroomable, whereas the people we were talking to were clean as cans. Only one office, of the eight we visited, looked like a place where a man was at work; this was Michael Mann's office, and the man at work was the director himself.

Our last screen meeting ended about midafternoon on a Friday—I dropped Diana at the airport and drifted indecisively over toward the 405, uncertain as to whether I wanted to go north to the 10 or south to the 605. I had decided to drive to Albuquerque on the fabled 40, but the 40 begins at Barstow, California, some 120 miles from LAX. It was a sunny afternoon in late spring; rush hour traffic was already in full flood and would stay at full flood for several hours. What I knew from a glance at the freeway was that any attempt to hurry would be foolish. Rush hour traffic on Friday in L.A. is an elemental thing—there was nothing to do but accept it, as I had accepted the eastern rainstorms that had delayed my flight that Sunday afternoon in Dallas.

Ontario, California, where I could pick up I-15 and proceed to

Barstow, was about forty-five miles away down a road along which several million cars were even then inching. At such times—and only at such times—I think of the Buddha and try to assume his patience.

From my point of view, though, being in a traffic jam on the L.A. freeways at rush hour has a definite edge, as human experiences go, over meetings with movie executives. It is the executive who decides to "take" the meeting, plucking a screenwriter from a large pool of supplicants as Nero might have plucked an apricot or a plum from a basket of fruit. As India has its untouchables, so Hollywood has its untakables, human fruit so spoiled by failure or treachery that no executive is likely to accept it.

Relaxed considerably by the happy thought that no more meetings would need to be taken for a while, I rolled up Lincoln Boulevard into Santa Monica and picked up the 10 right off Ocean, with the blue Pacific behind me. Ahead of me, downtown L.A. was also blue—only this was the blue-white of a hot smoggy haze rather than the blue sheen on the great waters.

In choosing the 10 as my avenue of departure I had accepted one huge risk, which was that I might have to spend the rest of the afternoon trying to get past the 10's frequently torturous juncture with the 5 (or Golden State Freeway), which occurs a few miles east of downtown. This exchange is tricky at the best of times, and midafternoon on a Friday is rarely the best of times. A congeries of freeways crosses that decisive area. The 710 comes pouring in from the west, and the driver who is not correctly positioned risks being sucked north or pushed onto the Pomona Freeway. This interchange, where so many roads go so many ways, is particularly vulnerable to small accidents on adjacent highways, whether these occur north, south, east, or west.

In this instance an accident on Colorado Boulevard, to the north, had this interchange hugely snarled. Many thousands of desperate drivers were struggling grimly to hold their lanes, or else signaled desperately to be allowed to slip into another before it was too late.

Movement could still just be achieved, but only very slow movement. It took me an hour and a half to get from the Santa Monica pier past the crossing of the 10 and the 5. Even so, I wouldn't say it was L.A. traffic at its worst. Once, years before, I had been on the 405, bound for the airport, when the whole system locked. A large truck turned over on the 10, near the Harbor Freeway exchange. Within a few minutes the effect was felt all the way across the basin. Soon, nothing moved. Thousands of canny drivers tried to slide off on the secondary roads, which soon locked too. The traffic strategists tore their hair, but for a time, nobody moved.

The little jam I was in as I inched toward the 5 was nothing so dramatic—it was just Friday afternoon traffic, a level of congestion that millions experience every week, if not every day, and it says something for the appeal of L.A. that millions put up with it without too much complaint, just because they like southern California.

I like it too. Jack Kerouac said long ago that L.A. was the west coast's one and only golden town, and he was right. Even though the movie executives have gotten much younger, the meetings even more vacuous, and the secondhand bookstores less numerous, I still get a lift from going there, still enjoy its confusion, its color, its brassiness. Soon, crossing Lower Azusa, I realized that I had missed lunch, so I slipped off onto Glendora Avenue and had a milk shake. There had once been lemon orchards near the place where I had my milk shake—all that pastoral wealth had been swept away by the relentless need of millions of Americans to nudge a little closer to L.A.

This need is understandable. For a century now the movies have been stirring American imaginings. Images created by the movies are as thick in the American consciousness now as the lemon trees once were in Glendora. Those screen images are so potent, and have been potent for so long, that one would think the reality of the big sloppy city where they were created would disappoint—but it doesn't. Of American cities only New York has an equivalent force. In many respects New York has a greater force, but then again, New York doesn't have Valet Park.

Personally, I've loved L.A. most as a book town. When I first began to scout it, at the beginning of the sixties, there were almost 150 secondhand-book shops in the greater Los Angeles area. Many of them were very exciting bookshops. Part of my impatience with screenwriting was that the lengthy story conferences kept me out of the bookshops for too much of the day. Fortunately, in those days, there was always Hollywood Boulevard, where many of the bookshops stayed open until midnight. For a young book scout, a town where the stores stayed open until midnight was synonymous with paradise.

Lower Azusa on a hot, smoggy, traffic-clotted Friday afternoon didn't seem quite as paradisaical, but I still liked it. In an essay written long ago I called Hollywood an ego zoo, and I still think the concept is apt. The demise of the great moguls has diluted the weirdness a little, but patches of vivid human color are still there, more or less wherever one looks.

Where writers are concerned, Hollywood is like the husband

who resents his wife because he needs her. Many writers have been made uncomfortable by this resentment, but I haven't. Like the traffic, it's just part of the price of L.A. George Bernard Shaw reportedly once made a famous remark to Sam Goldwyn. "The trouble, Mr. Goldwyn," Shaw said, "is that you are only interested in art and I'm only interested in money." I can endorse that sentiment. Writers in Hollywood are perfectly free to pursue their art, if they want to: they just aren't free to pursue it in movies, which, properly speaking, aren't their art anyway. If they think they're going to achieve much art by writing screenplays, they're barking up the wrong tree, and the louder they bark, the more they're likely to annoy the orchard keepers.

Even though I had cleared the 5, travel on the 10 was still pretty intense. The first forty-five miles of my travels took almost three hours. There's always a slowdown through Pomona, where traffic from the 210 spills in, but this afternoon the slowdown was *really* slow. Even when I finally passed Ontario and moved to I-15, traffic was still sluggish.

All this changed once I crossed Cajon Pass; "sluggish" was not a word that would be called to mind again on this trip. The 15 is the road to Las Vegas, where the weekend was just beginning. Las Vegas was some two hundred miles away, but many of my companions on the 15 seemed intent on making it in a couple of speedy hours—from the look of them as they whizzed by me they weren't rushing off to the Bellagio Hotel to see the famous art collection, either. I suspected they were hot for the slots, or anything else they could find. From the Cajon Pass to Barstow I drove about eighty-five, but cars shot past me at such high speeds that I became nervous about pulling into the left lane even when I needed to pass a truck. I didn't want to be rear-ended by some sport headed for the glitter at speeds above 110.

As I crossed Cajon Pass I noticed an intensification of the outlet mall concept: a sign said, *Neiman Marcus, 159 Miles.* A moment later another announced, *Williams Sonoma, 154 Miles.* At prevailing rates of speed these stores were less than two hours away. But how many low-flying pilots, Las Vegas in their sights, can hold Neiman Marcus in mind for 159 miles? Somewhere south of Vegas an oasis of goods must be accumulating in the desert. There's even a kind of outlet mall city just west of Barstow, though it didn't have Neiman's in it.

The only town of note between Ontario and Barstow is Victorville—during the silents era Victorville was famous as the Home of the Western. William S. Hart made his first western there in 1914, and hundreds of others were filmed nearby, in the spacious deserts. An itinerant fruit picker from Cincinnati named Leonard Slye changed his name a couple of times before he hit on Roy Rogers—the moniker that worked. The Roy Rogers and Dale Evans Museum is in Victorville. Ancients who still remember the westerns Roy made in the forties can go by and see Trigger, stuffed. To the north of town, in the desert across which so many hooves once thundered, there is a Dale Evans Parkway—Dale was Roy's wife. So far real estate along Dale Evans Parkway does not seem to have boomed, though a tentative little neighborhood is beginning to sprout to the south.

The sun drops very slowly this time of year—I'm hoping to be past Needles before it gets dark, even though I'm only driving a sedate eighty and cautiously hug the slow lane.

The need for caution diminished after Barstow, where I left

the fast 15. The trucks and I went east, onto the 40, and everyone else kept blazing on north, toward Las Vegas. Barstow is a bare-dirt community—no one has yet been foolish enough to try to irrigate this part of the Mojave. I am no sooner on the 40 than a sign announces that it is 2,554 miles to the other end of it, at Wilmington, North Carolina. At one time or another I've driven most of those 2,554 miles, but I don't intend to go that far with the 40 this time. Though it crosses much beautiful country, it's not a highway I love, perhaps because it's now essentially a truck route. There's scarcely a one of those 2,554 miles that won't have a dozen trucks on it at any given time. The need to keep a close eye on these rolling behemoths distracts one from paying much attention to the changing landscape.

From Barstow across Arizona, New Mexico, the Texas pan-handle, and much of Oklahoma, the 40 parallels old route 66. I head on over toward Needles; in the slanting light, the old road is just a few yards away. Frequent signs announce its proximity, but I don't see any tourists stopping to have their picture made on it. In fact, I don't see anyone pay any attention to it at all. Once past Barstow, the lure of Vegas removed, most tourists have their sights set on the Grand Canyon—the Big Crack to the northeast. An old road that no longer goes anywhere doesn't interest them.

Needles (on the 40), Blythe (on the 10), and Yuma (on the 8) are the three gritty, hard-bitten desert towns on the Colorado River. None of them are pleasant places. In the early sixties I had a car breakdown in Needles and had to linger for two days, wait-ing for a part to arrive by Greyhound bus. Never have I been hap-pier to see a bus arrive. The only pleasant thing to do in any of the three towns is to sit on the banks of the Colorado and watch the green water flow on toward Mexico. But there are so many desert rats and desperate drifters in these towns that one is seldom allowed to enjoy the river in peace. You may not get murdered or mugged (although some have), but at the very least, you'll get panhandled.

I slipped through Needles just at dark and decided to spend the night in Kingman, Arizona, fifty miles farther up the road. I'm out of California and into High Arizona now, a region not notably gentler, for stranger or resident, than High Albania (so called by the early Balkan traveler Edith Durham, in a famous book of that name). Mountains never seem to breed tolerance; they breed suspicion instead. Timothy McVeigh hung out near Kingman for a while, nursing his grievances against the government and learning to make bombs. Wovoka, the Paiute prophet who sponsored and preached the Ghost Dance, didn't die until 1932, a fact I mention because one of the last Ghost Dance uprisings occurred near Kingman. The Ghost Dance, a harmless form of Native American millenarianism, made the white authorities so nervous that they always overreacted in their efforts to suppress it, most notably at Wounded Knee, less famously at Kingman.

The town of Kingman itself is just a stop on the road—a rather harsh stop. In the hills around it are plenty of holing-up places, ideal for the resentful and disaffected—soldiers, as McVeigh was, of many strange gods.

Ahead of me, when I pulled back onto the 40 next morning, was one of the most beautiful high-country drives in America. The beauty of the high desert is nowhere better revealed than along the 40 from Kingman to Albuquerque, despite which I couldn't summon much excitement. I had felt more attentive in the Friday afternoon traffic, coming out of L.A. For me, L.A. is fun and the 40 a grind. Perhaps its the trucks, which were lumbering reluctantly upward as we rose, near Flagstaff, to more than seven thousand feet. The trucks reminded me of Hannibal's elephants; in the mountains their weight was burdensome to them. Now and then one truck would attempt to pass another, only to discover, once in the passing lane, that it didn't quite have the juice. I had a feeling that truck behavior was going to dominate my day, as it usually does on the 40.

Once past Flagstaff, onto the high desert plain, my spirits lifted a little. I was in the land of the Navaho now, and would be in it for much of the day. In polite discourse it's no longer proper to refer to "Indians"—"Native American" is the correct term—but along the 40 such strictures don't apply. Every few miles there's a big sign for the next Indian market, where one can acquire Indian jewelry, Indian blankets, Indian pots. Along this road "Native American" just wouldn't cut it—not commercially. If the new term were to be widely adopted, hundreds of roadside signs would have to be changed.

To the north, as I proceed along toward Gallup, lie the mesas where the reticent Hopi exist in the purity of their great space. One of my favorite small roads is the lovely, very lonely highway 87, which leads across the floor of the great space to Second Mesa, near Walpi, one of the most strikingly situated communities in America. I'm saving Hopi, though. I hope to come back down to it a little later in the year, crossing Monument Valley in the process.

I had vaguely planned to pay a visit to the Meteor Crater that morning. I have been to the crater several times and have always lamented that this mini Armageddon, this collision between asteroid and planet, occurred so long ago that no one—except possibly a few critters—was around to see it. Probably it was an event with no witnesses, like Bishop Berkeley's falling tree. Of course, if anyone *had* been there to see it they would immediately have been obliterated and could have made no report. Even more than the Grand Canyon—not far away—the crater is a place that owes nothing to humans. The Grand Canyon, in actuality abuzz with humans now, renders these humans antlike by its vastness; but the Meteor Crater is prehuman in a different, more impersonal way. It's one of the few terrestrial places where the vast galactic past reveals itself: cold, remote, humanless. It feels not only prehuman: it feels prelife.

This morning, though, when I start to pull off at the Meteor Crater exit, I'm discouraged to see at least twenty-five RVs ahead of me at the stop sign. The humans are already swarming over this prehuman place. The best time to be at the crater is early morning, before the swarm descends. I find that I have no desire to inch along behind all those RVs, so I pass on toward Albuquerque. I soon passed the Painted Desert, the Petrified Forest, and other tourist magnets. When I was six I made a trip with my father and mother to the Grand Canyon—it was the only trip I ever made with them in my life. I don't remember a thing about the Petrified Forest, but I know we stopped there because my father bought a ring made of petrified wood, which he wore for the rest of his life. Perhaps the petrification occurred when that meteor whopped into what is now Arizona. My only memory of that trip is of a fever-induced hallucination I experienced as we were on our way home. My fever had soared to 106, I'm told:

when I looked at the long, long stretch of route 66 ahead of us I thought the road was going up into the sky. The hallucination was so powerful that I remember it to this day: I thought my father was driving us into the sky, perhaps meaning to drop me off in heaven. I was too weirded out from the fever to be particularly alarmed, but I certainly was amazed.

In fact that illusion—of the road far ahead slanting up into the sky—is still available, even to the unfevered, on the 40 west of Albuquerque today. Once past the pueblos of Acoma and Laguna (the latter the home of the writer Leslie Marmon Silko), the road ahead is visible for such a great distance that it reproduces my sense that I'm driving into the sky. Dorothea Lange has a photograph of a long skinny road in Kansas that gives something of the same effect: only sky is at the end of that road. Thirty miles west of Albuquerque the 40 seems to rise into air, and it seems fitting that it should do this, for the sky here is so vast that it could subsume all things.

Coronado came past these pueblos as he sought the cities of gold, which means that the Indians of this region have experienced an unusually long colonial oppression. Acoma, the sky city built on top of a 365-foot bluff, revolted in 1599 and killed a party of tax collectors sent by Governor Juan de Oñate, who proved to be a revengeful man. He overwhelmed the Acomas, took several hundred prisoners, and cut one foot off any male over twenty years old, probably raking in a lot of seventeen- and eighteen-year-old feet in the process.

I have spent some time at Laguna Pueblo, with Leslie Marmon Silko and her father, Lee; I've been to Acoma many times, where the concessionaires are—to put it mildly—not friendly;

and I've visited, at one time or another, most of the pueblos near Albuquerque. I'm not comfortable there and am even less comfortable in the communities north of Santa Fe. These are all places where the troubles are old and the troubles are deep. The plains below the Sangre de Christo may be supremely attractive visually, as they were to Miss O'Keeffe, but socially they are very uncomfortable—the result of that long oppression. North of Santa Fe is where the toughest of the Indians and of the Spaniards survived. It's not a good place to have a car break down—not if you're an Anglo.

Albuquerque, though, is a nice small city, easy to enjoy. The Mexican food in Albuquerque—which is really Indian food—is as good as it gets, thanks in part to the rich chili fields near Hatch, New Mexico, just down the road.

I had meant to drive some fifty miles past Albuquerque and turn south off the 40 at Clines Corners, securing myself a lonely drive across the Staked Plain to Clovis. The route would have taken me through the town of Fort Sumner, where Billy the Kid is buried.

But the 40, as so often happens, had ground me down. I didn't want any more of those trucks. I'm driving these roads for fun—my scenario, insofar as I have one, leaves ample room for whim. I was in downtown Albuquerque, almost at the juncture of the 40 and the 25, when I surprised myself, swung south on the 25, and surrendered my rental car at the Albuquerque airport, trusting that a great silver bird could be found to waft me down to Dallas.

It was a good decision. The Albuquerque airport, newly spruced up, is one of the few airports in the world where a couple of hours' wait is a pleasure. It retains something of the airy simplicity of the old Fred Harvey Airport, while being coolly modern. Though I had eaten at two renowned restaurants in Hollywood, the bowl of green chili soup I had at a kiosk in the airport was the best food I had on my trip. The chairs in the waiting lounges are even comfortable, a great rarity in airports.

When we took off I scanned the ground below, hoping to spot the depression—I believe it's somewhere on a golf course near the airport—where a poorly secured hydrogen bomb rolled out of a military aircraft and fell to earth in the late fifties, a fact the military didn't choose to disclose for some thirty years. Fortunately this vagrant H-bomb didn't explode. If it had there would have been no more green chili soup in northern New Mexico for a while—and no Albuquerque, either.

J U N E

The 75 from Tampa to Miami. U.S. 1 Through the Florida Keys. The 41 West to Naples

THE FLORIDA KEYS, modest pods of earth protruding from a green sea, with U.S. 1 superimposed upon them, would certainly seem to validate my theory that littorals produce a gentle but distinctive seediness. The Keys constitute a kind of double littoral, the Atlantic on one side, the Gulf on the other. The little communities spread over these pods of earth provide more than enough seed to satisfy even the greediest eye.

My personal favorite is Key Largo—few communities its size can boast as many bright yellow buildings. The Banana Hill Center for the Healing Arts is perhaps the most vivid of these, though Flea Largo (an antique shop) is also splendid. For those who prefer pink buildings there's the Pink Juntique (another antique shop) just down the road. Opportunities to heal body, spirit, and mind—assuming these can be kept distinct—abound. Besides the Banana Hill Center, Key Largo offers several massage therapists and a European psychic; nearby Islamorada boasts a mentalist.

Deep in the mind's eye, as one drives past these brilliantly colored establishments, are black-and-white memories of the young Lauren Bacall and the not so young Humphrey Bogart. They made *Key Largo* together in 1948.

The drive down the Keys, from Key Largo at the north end to Key West at the south, is a good place to start thinking about oceans. Two are frequently visible from U.S. 1. Plains of water stretch away both left and right. But lots of people live in the Keys, and lots more visit—the 125-mile drive down the 1 is apt to be a slow experience. Once in a while a drawbridge has to be waited out. The farther south I go the more varicolored the sea becomes. Here and there one sees patches of vivid green, as distinct as farm plots in Iowa, though elsewhere the sea is gray, metallic, or blue; in part these variations result from the ever changing light.

When I had tried to come down the 75 earlier in the year, the Everglades had been burning, so I waited. To reach the Keys I flew to Tampa and crossed the state on the southern end of the 75. The Everglades were no longer aflame—I wanted to pass as close to them as I could. All across the state the roadside foliage was verdant and bright, a medley of deep greens. Stopping at one point to relieve myself I found I could have used a machete, just to get deep enough in the foliage to be out of sight of the road. The plants in their millions didn't welcome invaders—not even momentary invaders. From Tampa this kingdom of plants extends south more than two hundred miles. Highway 75, the Everglades Parkway, is also known as Alligator Alley, though the actual alligators have been expensively fenced off the road.

The question that rose to mind as I studied this fortress of palmetto is why Ponce de León, or Narváez, or de Soto or any of the European explorers bothered to hack their way onto this peninsula. Surely they could not have supposed for long that the treasures of the Orient were hidden in these marshes. But they came and kept coming, determined in their greed to have this new continent even if there didn't immediately seem to be much in it worth wanting. They rooted out the native peoples and pressed on. Soon I was passing through the Miccosukee Reservation—the Miccosukees are now the proud owners of a new casino. They are also deeply involved in the rapidly intensifying debate over how to save the water-deprived Everglades. I feel, as I drive through the Big Cypress National Preserve, that I'm looking at a kind of low rain forest, where the natural dramas occur in the marsh rather than in the canopy.

Once past the Big Cypress, the marshy south Florida plain stretches away to the Atlantic—very beautiful. Twice, crossing bridges over small inlets as I proceed down into the Keys, I have to pull left in order to avoid rear-ending pelicans, which seem to claim the slow lane for themselves. Perhaps there are delicate thermals over the slow lane that motorists are unaware of—but the pelicans know.

In the lower Keys the brilliance of the foliage, so evident all across Florida, begins to bleach out, as if the sea and the winds have conspired to steal the nutrients that made the plants so bright. By the time one reaches Deer Key the scrub seems rather dingy. On this key the traffic is held to a strict forty-five miles an hour, in hopes of sparing the rare Key deer. A sign informs us that fifty-five Key deer have been killed by motorists already this year. We all inch along, but I see no Key deer.

As for fabled Key West itself, it might be a blessing if Disney just bought it—it's essentially a theme park now anyway, and Disney would at least know how to manage the parking, which is, at present, a torture. Certain historic districts have been more or less roped off, in hopes that visitors will intuit from them the easygoing charm for which Key West was evidently once noted. This takes more intuition than I can summon; "easygoing" is the last adjective one would apply to Key West now.

Though my drives along America's roads were not designed as pilgrimages to the homes of writers, I felt it would be churlish, since I was in Key West, not to see Hemingway's house. I had looked at it once long before but had not gone inside. In northern Michigan I had been in the vicinity of the fishing camps of his youth—the Big Two-Hearted River country—so it seemed only fair to check out his place in Key West, from which he also did a good bit of fishing. I found the house, handed over $8, and took the tour, an experience that proved, on the whole, disquieting.

As a writer-bookman I've always been curious about writers' libraries—what books they read, what books they keep, what books they *don't* have. I have seen the library of Evelyn Waugh in the University of Texas at Austin; I've also, at the University of Tulsa, seen Edmund Wilson's books, and Cyril Connolly's. Hemingway read—he had a large library. There is a photograph of him standing on a small library ladder, arranging the books in his villa in Havana, La Finca Vigia. There are nine thousand books there;

from the look of the shelves in that photograph they are good books, well arranged by their owner and principal reader.

But the books in the Hemingway house in Key West are mostly just moldy junk, shelf fillers, worthless books of the sort that can be gathered up at any Goodwill store. It's hard to believe Hemingway had any hand in their selection. Though the house itself is gracious, well shaded, and spatially comfortable, the furniture and *objets* seem ill assembled, indifferently arranged, mediocre. It's hard to imagine Ernest Hemingway being really comfortable in these rooms; there's a stiffness here that's a little off-putting. In the late twenties he first showed up here with his second wife, Pauline Pfeiffer. There were two wives yet to come, Martha Gellhorn and Mary Welsh. Any of the three, I suppose, could have been responsible for the furniture in the Key West house, with its nightclub-era feel.

I didn't want to engage with Miami on this trip—Miami, for me, is heavy work. I went back through the Keys and, once on the mainland, turned west on highway 41. My hope was to get a slightly more intimate look at the Everglades than had been possible along the 75. Highway 41 goes all the way across the peninsula, from the Atlantic at Miami to Naples on the Gulf. The largest town along this highway is Ochopee, and it is not a large town. An observation tower in the southern part of the Miccosukee Reservation does allow one to look deep into the Everglades—the vista over this immense, complex water world is beautiful, spooky, compelling. Saving the Everglades is going to be one of Vice President Gore's principal concerns—we are sure to hear a lot about it in the 2000 presidential campaign.

I want the Everglades saved, and I want to know them better, but on this particular afternoon, I can't quite get my mind off the Hemingway house in Key West. I had gone to it casually but left it puzzled and somewhat disquieted. It must be that I want to think of Hemingway's taste as being the equal of his best prose—that would be, in the main, the prose written before he ever set foot in Key West, which he did, I now gather, at the urging of John Dos Passos.

In my view the furnishings in the Key West house, as it sits today, are more suggestive of Hemingway's *worst* prose. As I drove across south Florida and finally turned north, toward Tampa, I began to wonder if I had invested more in Hemingway's early excellence as a writer than I had realized. I hadn't read him carefully in at least thirty years and had, in the main, avoided reading much about him, since much of what has been written about him makes him seem insufferable. Edmund Wilson mentions in his journals that Hemingway had become insufferable; and yet Wilson also records how shocked and saddened he was by Hemingway's suicide, mentioning what a loss he felt, for Hemingway had been one of the writers most important to Wilson's generation.

I was not exactly depressed by my visit to the Hemingway house—just disquieted. I chewed on this disquiet all the way across Florida and halfway up its length. The point may be that it's always tricky to go near writers whose work you really like. They may turn out to have bad furniture, or tacky women, or both. I once was a prolific reviewer of contemporary fiction, always for newspapers. What I discovered early on was that it was much easier to be generous to the work if one kept free of any association with the man or woman who wrote it. I once met a well-known writer in an airport; he recognized me and immediately struck up a conversation about his favorite subject, himself. Though a jerk, an asshole, and a bore, he was, and he continued to be, an excellent writer; but from then on, I avoided reviewing him. I couldn't forgive the books what I knew about their author.

The simplest analysis one can make about Florida is that its character, tone, flavor, ambiance is largely determined by one geographic fact—it's a peninsula, with most of its population balanced on the coastal rim. The interior of the state is comparatively empty; Orlando is the only large or largish city not set on a beach, unless you count Gainesville, which essentially is just a college. Thus almost all Floridians are people of the littorals, which may explain why so few of the men bother with long pants, or even, for that matter, shirts. Though I was only in Florida for a day I saw as many deeply bronzed bodies as I would expect to see in Arizona in a week. The women of Florida—the specter of skin cancer no doubt on their minds—are a good deal more decorous than the men.

Back home, after my short trip down the south end of the 75, I felt that the trip had raised questions that it would be fun to investigate. Did Ponce de León really believe that the fountain of youth lay somewhere out in the pines and the palmetto, when he anchored off the northeast coast of Florida in 1513? He had come to the new world in 1493 and was soon hard at work putting down native rebellions in Puerto Rico and elsewhere. He was not simpleminded, but he may have been at least somewhat beguiled by the prospect of rejuvenation in a marvelous fountain; many unsimpleminded people with the same desire keep the fashionable spas of the world humming even now.

Also, I wanted to read Henry Adams again, on the administrations of Jefferson and Madison, in the hope of understanding just how it was that we wrested Florida from France, even though, at the time, it was still claimed by Spain. And I certainly intend to break my ban and read a bit about Hemingway. I want to find out which wife, if any wife, was responsible for the furniture in that gracious house on Whitehead Street, in Key West, Florida.

JULY

Washington, D.C., to Dallas
via the 66, 81, 40, and the 30

WHEN I RETURN to Washington now I feel as if I'm returning to the place where I lived before I died.

That I have this feeling, a powerful feeling, takes some explaining, in view of the fact that I'm alive, healthy, active, and to all appearances, living an enviable life, mostly in Texas. But the feeling I get—that I was living in Washington when I died—is too potent to ignore. What I mean by it is that I lived there before my heart surgery, which occurred on the second of December 1991, at which point I experienced—if I may be permitted an oxymoron—a death that wasn't fatal: my body lived on but my personality died, or at least imploded, disintegrated, shattered into fragments. For the past eight years I've been struggling to collect and reassemble these drifting fragments of personality and I believe I have now reassembled most of them. They fit together a little crookedly still, but that's only to be expected after one has been radically cut open.

I don't intend to spend too much time on the mind-body ambiguities that this experience has exposed, but I do think it's worth pointing out that I'm far from being the only person to experience personality death while continuing physical life. Victims of stroke very often survive physically but lose all trace of personality—that which, in their view and the view of their loved ones, made them uniquely *them;* and soldiers, after combat, too often feel that their souls have fled while their bodies continue to function.

Of the little drives I've made so far along America's roads, this route from Washington to Archer City is the most clearly a retracing of an old, deep-worn, homeward path. During the twenty years in which I was mainly domiciled in Washington I drove this route at least three times a year, which means I've driven it about sixty times, often enough to have acquainted myself with virtually every gas station and convenience store along the way. Sometimes these drives had something of a utilitarian motive—I might be hauling books from the bookshop in Washington to the younger bookshop in Texas; but they were still, in the main, homing drives, long lopes down the highway that I took whenever I had had enough of the east.

As I prepared to drive those same overfamiliar roads again it occurred to me that my effort was obliquely Proustian, a retracing of my past that is analogous to the many rereadings I've done in the last few years, always of books I read before the surgery. In these rereadings and redrivings I'm searching, not for lost time, but for lost feelings, for the elements of my old personality that are still unaccounted for. I'm not anguished about these absentees, just curious and somewhat wistful. I don't really expect my old personality to be waiting for me at a rest stop in Tennessee, or a Waffle House in Arkansas, but I am still listening for chords I haven't heard in a while, wondering if a passage in a book or a place I once liked along the road will cause them to sound again.

I have not been in Washington much, since my operation. In the first years after surgery I didn't know what had happened, or was happening, but I did know that I needed to be under western skies, bathed by western light. Personalitywise I was almost a blank. I had had my competitiveness removed, and a person without competitiveness has little chance in Washington, D.C., whose elite echelons—journalistic, legal, bureaucratic—comprise one of the most competitive social entities on earth. I realized I was in no shape for Washington, so I left, even though that meant deserting my bookselling partner, Marcia Carter; it also meant ceasing to have much part in the affairs of Booked Up, the exciting rare-book shop we had built up together over almost thirty years.

But I left—I couldn't help it. When I did reconnect with the antiquarian book trade, almost five years later, I did so by shifting most of the general stock of Booked Up (that is, the cheap books) to Texas, where I could be a bookseller and yet be within the embrace of those skies.

I suspect that another factor in my current bittersweet response to Washington is the you-can't-go-home-again problem, though in my case it's not so much a matter of going home as merely going *back*. Washington never felt like home, but it was nevertheless a place where I worked effectively for some twenty years. Marcia Carter and I had great fun hustling rare books in the Washington of the seventies and eighties, when many of the substantial personal libraries of the capital were being broken up. The opportunity, which came to us almost as soon as we opened, to buy from the libraries of such legendary Washingtonians as Huntington Cairns, James M. Cain, David Bruce, and Alice Roosevelt Longworth was, to young booksellers, extremely exciting. The few antiquarian booksellers operating when we opened in 1971 were old and tired; for a few glorious years we had the libraries of the town almost to ourselves; we bought wonderful books, established an exceptional stock, had a great run.

Part of the trick of being happy is a refusal to allow oneself to become too nostalgic for the heady triumphs of one's youth. Pickings are still good in Washington, and Marcia is still there, picking them, but they are never likely to be quite as good as they were in our first years, which is perhaps what causes the you-can't-go-back feeling to assail me when I land in Washington now.

David Streitfeld, for twenty years the preeminent interviewer of literary folk for the *Washington Post,* recently made me his five hundredth interviewee (Wilfred Thesiger was his 501st).

In his ruminations on my life and character David quotes some friend of mine as saying I was a good hater, a comment that produced indignation in the bosoms of a number of my friends, some of whom don't consider me any sort of hater at all. In the course of arguing this point I realized that I once *did* hate something: that is, eastern privilege, or what Lyndon Johnson simply called "the Harvards." In Texas the only Harvards I had known were one or two courteous old professors at Rice—I didn't, at the time, know what the president meant. It was evident, even in Texas, that some people had money and some people didn't, but I never really associated this disparity with class: out west there were only two classes, middle and working. The first thing Washington revealed to me was the existence of an American upper class, possessing qualities, opportunities, and privileges that I hadn't realized existed. I had been blind to the existence of such a class, and such privileges, but I immediately realized why Lyndon Johnson and, later, Robert Dole hated and distrusted that class.

Johnson and Dole were, of course, politicians, but I was a novelist of manners, to whom social stratification is as vital as lifeblood. Much as I resented eastern privilege I immediately realized that it was teaching me something I needed to know: what would the novel of manners have been without the struggle between old money and new money? In *Terms of Endearment,* begun only a year or two after I moved east, my heroine Aurora Greenway is class proud enough to claim Boston ancestry, although she is really only from New Haven.

By the time I had lived in Washington twenty years I had calmed down a little about eastern privilege, in part because it had become evident that the established class was losing its hegemony and had become vulnerable, as established classes have always been, to superior energies from below. But I was shocked and still am by the desperate ambition of Washington's social elite, or at least by the part of it I was exposed to, which was largely the journalistic elite. Nowhere is the Darwinian struggle more bloodily evident than among the princes and princesses of the press, whose hapless children are forced at age two or three to start ascending a formidable ladder of schools: Beauvoir, then St. Albans or Sidwell Friends, then Harvard, Yale, Princeton, or Brown, where, to uphold the family colors, they must not only get in but secure *early acceptance*, the sine qua non of survivability in that peculiar social gene pool.

Looking back on my twenty years in the capital I now realized that, though I adjusted to Washington socially and even sat at the dinner tables of such social exemplars as Joseph Alsop and Evangeline Bruce, there was a more basic need that kept forcing me to drive away: the small eastern sky. Washington *has* sky, of course, but it doesn't have nearly enough to still my yearning for the plains. I felt claustrophobic in the east and stayed there mainly because the book buying was so exciting. Even so there were many times when, Huck Finn–like, I simply lit out for the territory, to the place where the sky swelled out.

My first shock, on this visit, was the resplendent new Reagan Airport, a clean-as-a-pin new facility which reminded me immediately of the shopping mall at the intersection of Westwood and West Pico in L.A. National Airport, which was what this one was called before it was renamed for Ronald Reagan, had long been one of the grubbiest transit points in the nation; it was usually full of frazzled congressmen and their even more frazzled aides, struggling to get back to their districts for a hard weekend of fund-raising. National was also filled with the underclass. The new Reagan version is so spiffy the congressmen probably stop and straighten their ties before flying off to the fund-raisers. As for the underclass, it's gone, probably driven back into the streets by the fact of so much cleanliness.

I spent a pleasant day at the Washington Booked Up, picking a few plums to take home to Texas. By the end of the day I had already accumulated $60 worth of traffic tickets. Municipal Washington seems still to be financed largely by parking fines, one thing that hasn't changed since my day.

What *has* changed is that the great social stalwarts of the seventies and eighties are now dead. Evangeline Bruce, last of the great Washington *salonnières*, finished her book on Napoleon and Josephine, suddenly went blind, and then dropped dead. Pamela Harriman was stricken while swimming in the pool at the Ritz, in Paris. Joe Alsop finally succumbed to lung cancer. Clark Clifford, that old smoothie, slipped seriously at the end of his life, mismanaging a lot of Pamela Harriman's money but a lot of bank funds as well; he who sat at the right hand of several presidents died more or less disgraced. The title of Joe Alsop's autobiography, *I've Seen the Best of It*, in which he describes not only his life but the era of the Wasp ascendancy, as he liked to call it, was accurate. He *had* seen the best of it, and so had Evangeline Bruce, Pamela Harriman, and the rest of the old patrician crew. Katharine Graham, Paul Nitze, and a few others of that class are still with us, but not by much. Greatly privileged snobs though they all were, they were yet a great deal more interesting than the mediacrats who have succeeded them.

I slipped out of Georgetown early, before the parking patrol got going—though I was not at first sure which route I intended to follow. Sometimes the desire to see both ends of a road nags at me. I considered going down to Wilmington, North Carolina, where I-40 begins, and following it west for some of the 2,554 miles that lie between Wilmington and Barstow, California, where the 40 ends. Two things deterred me, the first being the fact that if I dropped down to Wilmington I would have to spend most of a day on I-95, the road that no one loves. Also, I had heard that much of the eastern part of the 40 was under repair, which is not surprising. In the course of the highway's long marriage to the big trucks, spousal abuse has taken its toll. To the cross-country driver, road repair is a serious irritant. One expects to be in traffic jams on the Harbor Freeway, or the Washington Beltway, or the Long Island Expressway, but that's different from being stopped dead still for twenty minutes while trying to get past Cookeville, Tennessee.

I was prepared for a few delays on the 40, but found that I couldn't face the prospect of these delays occurring in the Triangle, as the area around Raleigh, Durham, and Chapel Hill is now called. This prosperous area is the homeland of the yuppie redneck, a southerner with working-class prejudices and upper-middle-class money; the Triangle at almost any time of day is apt to be one long snarl of affluent traffic. I decided, with no reluctance, just to skip it; so far as landscape goes there's little to see along that route that can't be seen anyway in southern Virginia or southeastern Tennessee.

This decision left me free to follow the old road home, across the Key Bridge and into Rosslyn, a community that seems to consist only of a few tall office buildings, one of them the gleaming tower from which emanates *USA Today*.

Up the Potomac a little distance I saw, as I crossed the bridge, the three protruding rocks known to locals as the Three Sisters. When I first moved to D.C. in 1969 these three rocks were under threat—road rage among commuters had already risen to such a dangerous level that the authorities considered punching a superhighway right through Arlington and across the river, obliterating the Three Sisters. The first political protests I took part in after coming east were the fight to save these three rocks, a more or less leisure-time activity the city's activists devoted themselves to when not staging more serious protests against the Vietnam War. I-66, which I soon join, did get built but didn't obliterate the Three Sisters. A few towels were spread on the largest of the rocks, and two swimmers were splashing around.

Soon, outbound on the 66, I was able to study a congealed mass of commuters stuck behind an accident on the other side of the road. This was the secretariat, the ant people, inching their way in from well-managed colonies in Falls Church, Vienna, or Fairfax to keep the government going. These commuters were obviously seasoned ants, neither surprised nor particularly dismayed by the delay they were experiencing—after all, being at work in one of the great gray buildings where America gets administered is probably not much more exciting or even particularly different from being on a freeway in a parked car.

Very soon I was beyond the Beltway, the trafficky circumferential that manages to give Washington proper something of the aspect of a walled city. The Beltway protects our governing body from whatever chance infection of common sense might occasionally waft in from the country at large.

While flying east I had leafed through *Time* and *Newsweek*, both of which, that week, had chosen to confront the specter of urban sprawl. Evidently urban sprawl has replaced the ozone hole as the Worry of the Week. A drive past Fairfax, Reston, and Manassas will allow any citizen a good look at urban sprawl in its most rampant manifestation. In Fairfax I noted that the embattled National Rifle Association (NRA) has a splendid new building from which to promote the universal availability of assault rifles. The glass on the new building is only a little brighter than gunmetal blue.

Manassas, just down the road, prides itself on being the city that turned back Disney. Michael Eisner and company had really wanted to build a historical theme park near the battlefield, but the city rose up and Disney, in the end, backed off. I'm not sure how much this victory had to do with the folks in Manassas really wanting to preserve the historical character of this now-suburban community. It may have just been that the citizens realized that anything that would bring more traffic into the area—such as a vast theme park—would end up immobilizing them all completely and, in the end, driving most of them mad.

I-66 is a very short road, extending only from Rosslyn to I-81, an hour's drive at most. It allows workers from Fairfax, Reston, and the Dulles Airport area to slip more or less quickly inside the Beltway and help keep the country running. Thanks to the 66 it is possible for some people to live a bucolic life in or near the Blue Ridge Mountains and yet still work in Washington.

I had not been out this road in some years—urban sprawl had obviously made a good showing in my absence. Vast malls, undreamt of when I was last there, are now plentiful in the outlying suburbs. For a few miles I thought that urban sprawl might now be sprawling all the way to the Blue Ridge, but that had not occurred. To reach the Blue Ridge, the malls and the myriad housing developments that support them would have to surge over the splendid horse farms of the gentry, out near Middleburg, Warrenton, Upperville, and The Plains. This area is one of the great old-money country-house enclaves still left in America, and the gentry who live there have been trained from birth to deflect just such irritants as urban sprawl. What it takes is the ability to commit more money to preserving their well-manicured purlieus in Loudoun and Fauquier Counties than the developers can commit to destroying them. In most parts of America some deference is paid to landed wealth, but exceptional deference is shown the horse farm set in Virginia—after all, the horse farm set has been there ever since there was a Virginia. Visit Mount Vernon, Monticello, or Oak Hill (James Madison's home) and you'll see what I mean. In Loudoun and Fauquier Counties the squirearchical habits of the English gentry, including shoots and riding to hounds, have come down intact through the centuries.

I-66 slips discreetly through the region of the great estates, none of which are quite visible from the road. Paul Mellon, art's First Patron, lived near Middleburg; Forrest Mars, of the candy fortune, had an estate near Warrenton, not far from the famous Foxcroft School. Since squires rarely wash their own dishes, do their own laundry, or mow their own lawns, this area is one of the last strongholds of the "domestic," as the term would have been understood in England until recently. Quite a few humble families, who look as if they had been created just to sing Child's ballads, spend their whole lives on these estates. Since the squires and their ladies, from having so much money, often lapse into adulteries, addictions, or other forms of modern misbehavior, it is often the domestics who provide such stability as the families enjoy.

Loudoun and Fauquier Counties are unusual in that so much of what remains of the country-house gentry is concentrated in them, but they are, after all, just two counties. The casual traveler on I-66 would never know they were there. I remember them mainly for the books I know their noble owners have casually gathered in—just beyond Front Royal, for example, I passed a modest country house in which resides a fine copy of *Paradise Lost,* with the first title page, now rare. It had belonged, I believe, to Lord d'Abernon, who had at one time had a crush on Anita Loos, whom I met at a party in Arlington, late in her life. We talked about screenwriting, which she had begun doing for Mr. Griffith in 1912. But Lord d'Abernon had other crushes, on one of whom he had bestowed the *Paradise Lost.*

The 81, when I hit it, was as packed with traffic as the Beltway had been. The drive down the great valley of Virginia, with the Blue Ridge Mountains to the left and the ranges of Appalachian West Virginia on the right, is one of the most beautiful drives in America—if what one is looking for is pastoral perfection. Urban sprawl *hadn't* jumped the mountains. The small towns in the Shenandoah valley are still just small towns. Virginia is in the grip of a great drought; the grass in the valley is burned to an umber color. This drive is one of the few eastern drives that I really like, but on this trip I have little opportunity to look at the scenery. The southbound 81 is as tight with traffic as any urban freeway. Usually it eases a little south of Roanoke, but this time it didn't; so, in four rapid hours, I'm through Bristol and into the long thin state of Tennessee. I had passed through Virginia without even taking my foot off the gas, just doing my best to not be run over by a truck.

In Tennessee I soon met my old nemesis, I-40, and discovered that the rumors of massive roadwork along it were not lies. The roadwork is a factor from Knoxville all the way to Little Rock. Knoxville was James Agee's hometown. I once had an intense Agee phase and had several times drifted around Knoxville, looking for the neighborhood described in his early lyrical prose idyll "Knoxville: Summer 1915." This time, before I could even get into central Knoxville, I was engulfed by a traffic jam of a scope and dimension to equal that I had experienced in L.A. a few weeks earlier. The converging streams of the 75 and the 40 were not converging smoothly just then. Things were so tight for a while that even a simple lane change had to be negotiated an inch at a time. I was almost to Oak Ridge before I could even ease out of the fast lane—and Oak Ridge is well west of the Agee neighborhood. Most of Agee hasn't worn especially well, but *Let Us Now Praise Famous Men* still seems to me a masterpiece, an inspired wedding of Agee's text and Walker Evans's photographs. Two highbrow, elitist young men, sent by *Fortune* magazine to do a story about sharecroppers, came back with more than either of them—or *Fortune*—had bargained for. The gloom of the south is there.

I had meant to stop somewhere between Knoxville and Nashville on this drive, but when the Knoxville traffic finally broke up there were still four seductive hours of soft Tennessee daylight to be enjoyed, not quite enough to take me all the way to Memphis but too much to waste sitting in a motel in Harriman, Tennessee, or somewhere.

The happiest surprise, along this route, was the discovery of a nice new bypass around Nashville, a mean town that I usually race through with my head down, hoping not to be shot, broadsided, or otherwise slowed. Off the freeway, Nashville traffic has always frightened me; in my view it's the city in America where red lights are the most likely to be ignored, often by speeding citizens who are speeding on something other than gasoline. Many of Nashville's drivers could fairly be said to be living in their own heads. Though a large, sophisticated city now, Nashville still seems to me to be dominated by the suspicious us-and-them psychology of the mountain hollows; Appalachian passivity alternates unpredictably with Appalachian ferocity.

Despite delays for construction, which were numerous, it was still soon evident to me that I was traveling this route much faster than I had ever traveled it on my sixty previous trips. Those trips were made in the era of the fifty-five-mile-per-hour speed limit. Like most long distance drivers I usually gamble on being able to get away with ten miles over the speed limit: thus fifty-five means sixty-five, and the seventy that now prevails means eighty. A driving day that under the old rules normally covered about six hundred miles now covers between seven and eight hundred. When I pulled up for the night in Jackson, Tennessee, I had come 780 miles; but for all that road repair, I could easily have made it another seventy-five miles and slept in Memphis, beside the Mississippi. Even at 9:15 P.M. the sky had just darkened enough above Jackson to reveal a few stars and a crescent moon as I walked into my motel.

I wanted to see the Mississippi at first light, and rose early to accomplish this, but the pleasure was somewhat diluted by the appearance, in Memphis, of a vast, shiny pyramid, which had been plopped down almost on the riverbank, north of the main bridge. Just when I had expected to glimpse the mighty river, there was a glass pyramid blocking my view—possibly it's a convention center, meant to remind folks of the other Memphis, the one in Egypt.

No building can really defile the beauty of that river, though. Sunlight was just beginning to burn away the mists along the bank, and a single barge pushed up the channel. My fellow travelers and I arched high above the river and descended into Arkansas. In the parking lots of the vast truck stops in West Memphis the great trucks were just beginning to belch and cough, as the truckers began their day.

Between Forrest City and Wheatley long fields stretch to the horizon, and I watched the sky become a western (or at least, a midwestern) sky again. It was here, on the Arkansas flats, that I would invariably feel a lifting of my spirits on earlier drives out of the east. Beneath these skies the horizon would regain its mystery and I would feel at home once more.

Like Nashville, Little Rock now has a nice new bypass that carries one over to I-30, the convenient road that connects Little Rock with Fort Worth. The bypass allows one to marvel at the splendors of downtown Little Rock without even having to slow down.

Long before I reach Texarkana I'm reminded of an old truth of the road, which is that Arkansas drivers, of all drivers in America, are the least likely to yield the fast lane. They just won't let you by. This is not because they are macho, it's because they're indifferent to even such modest subtleties of the road. To Arkansas drivers the lanes look the same and are treated the same. It's pointless to ride on their bumpers, as many frustrated travelers do. The rule in Arkansas seems to be that what's behind you needn't concern you—drivers there make only the most sparing use of their rearview mirrors. No wonder the Clintons blew out of this state—no one could accuse them of not knowing what to do in the fast lane.

In my earlier drives I very often spent the night in the president's hometown, Hope, Arkansas, a none-too-prosperous small town about thirty miles east of Texarkana. I'm pleased to see that Hope now has a Western Sizzlin'—in the old days the one disadvantage to staying there was that it was hard to buy even a hamburger after 8 P.M. Late arrivals sometimes had to make do with the vending machines at the all-night garage. Hope has since put up a nice sign, proclaiming it the hometown of President Bill Clinton—no pomposity, no William Jefferson Clinton, just plain Bill, a boy who blew out of there and is still blowing.

I remember only one fact of interest about Texarkana, which is that William F. Buckley Jr. has his limousines customized there—shaped and fitted to his exacting standards. If Mr. Buckley should decide to limber one of them up by driving west across the state to El Paso, he would be looking at a long stretch of highway—almost nine hundred miles.

THIS DRIVE along the 81, the 40, and the 30 turned out to be no fun. I realized, even before I got out of Virginia, that I would have done better to choose fresher roads. I probably decided to drive my old route yet again in hopes that it would tell me something about my state of mind in my Washington years, when I regularly drove this set of roads. In those years, of course, I lived in the east, and missed the west—not its peoples, but its landscapes. I was having a more stimulating time hustling rare books in Washington than I had been having as an academic in Houston—but I still missed the plains. Though I can live happily almost anywhere for a few months, my attachment to the plains landscape is irreplaceable. Like the loners of the Oklahoma panhandle, I don't need much society, but I do need a lot of sky. I enjoy driving around on the plains, but—as these essays testify—I live on them again and don't need to drive *to* them as I once did.

Viewed in that light, this trip was a small miscalculation, one that told me nothing about any previous state of mind. It just brought me home.

AUGUST

Short Roads to a Deep Place

I GREW UP in a dirt-road world. The great roads that I have been driving so blithely were not there in my early youth. Highway 281, a paved road at least, *was* there, but we didn't use it much. It seemed like a road for others, not for us.

Our ranch house was at the foot of the plains, a dot at the center of the great circle of the horizons. The forms of mobility available to me in childhood were ancient ones: the horse and the wagon. We had an automobile but used it only very occasionally.

Twice a week my grandfather William Jefferson McMurtry plopped me behind the saddle on his big horse and took me with him to Windthorst, a small town six miles away, to get the mail. We traveled cross-country, my grandfather taking care to close the several gates to neighboring pastures as we passed through. The horse was so big and slow that I could have danced a jig on his rump. He was reliable, but he wasn't fast, which didn't matter. We were in no hurry. Getting the mail usually took a morning; my grandfather would often linger in the general store, which then doubled as a post office. He liked to get the news as well as the mail.

The most exciting thing that happened on any of these trips was a freak hailstorm, which we rode out under a large oak tree. I watched, amazed, as the prairie was covered by bouncing white balls; the hail crunched like glass under the horse's feet as we rode home.

The pickup, as a ranch utility vehicle, did not become common in our part of the country until after World War II. They existed, but most smallholders were too poor to afford them. In times of severe drought, when it became necessary to feed the cattle a little something, we fed cottonseed meal out of a wagon drawn by two brown mules. The mules were in even less of a hurry than my grandfather's horse. Time was cheap, or seemed to be to everyone except my father, who was responsible for making ends meet in those Depression years.

I liked riding in the wagon, which was usually driven by one or another of the old cowboys who worked for us from time to time. Being in a wagon was much nicer than being on my detestable pony. My dog was constantly leaping out of the wagon to pursue jackrabbits, though the jackrabbits ran off and left him as easily as the Road Runner outdistanced Wile E. Coyote in the famous cartoons. In winter the sideboards of the wagon afforded some protection from the biting wind. The old cowboys eased the wagon slowly down the rutted roads to the feed grounds, and then eased it back over the same ruts to the barn.

Moving cattle to market by truck was something else not commonly done until after World War II. In the early forties it was still necessary to drive cattle to a railhead of some sort in order to get them to market. The nearest shipping point was fifteen miles west, at a place called Anarene—it may have once been a town, but by the time I came along it was just a set of corrals by the railroad, with a small general store across the road where thirsty young cowboys could purchase a Delaware Punch.

I was first allowed to help drive cattle to Anarene when I was four years old. My mother thought this endeavor premature, but then she has spent a lifetime thinking virtually all endeavors premature. My father, in this instance, prevailed. I was four—it was time I got on with being a cowboy, and besides, all I was going to be required to do was amble along behind some cattle for fifteen miles. We were driving the cattle along a dirt road toward a railroad track: the possibility of stampedes or other excitements did not exist.

I was thrilled to be allowed to go with the cowboys, but the thrill was mostly anticipatory. When we set out at dawn I was still wildly excited, but within about half a mile, my excitement gave way to tedium, both for myself and for the cattle, and this tedium lasted all day, as we plodded along the straight dusty road. The following year, because the Anarene pens were for some reason unavailable, we made the drive again, this time going all the way to Archer City, an extra three miles.

These were the tamest imaginable trail drives, and also the last trail drives. By the time I was six or seven, cattle trucks were common and we no longer drove cattle, except from pasture to pasture. When it came time for our cattle to be sold, big trucks came to the ranch and got them. A form of pastoral droving that had existed in America for more than a century abruptly died.

I mention these slow travels—horseback trips to get the mail, wagon trips to feed cattle, and short trail drives to the railroad—mainly to illustrate how dramatically the pace of travel has increased in my lifetime. For a few brief years in my childhood I experienced travel as something slow, which is how human beings had to experience it until about the middle of the nineteenth century. I went no faster than a horse could trot, or a wagon roll, of course not realizing that I was seeing the end of a very old tradition. Foot speed, horse speed, and wagon speed set the pace of human mobility for many centuries. Then came railroads, automobiles, airplanes; and the day arrived that I described in my previous chapter, when a big international airport bulged with people who were ready to cut their own throats because they had to wait an hour for an airplane to hurl them across the country, a thousand miles or more, in time for supper.

And as to cattle, I recently heard that a number of dairy cows had been FedExed to Russia; many an old trail driver would have goggled at that.

I myself, less than (I hope) half a lifetime after bouncing down our feed roads in a slow wagon, flew to Europe on the Concorde—I once went over and back in a single day, in about the length of time it had taken me to follow those sluggish cattle a few miles down the road to Anarene.

Today I live again in Archer City and drive quite often down the stretch of dirt road along which I dully but dutifully followed those cattle in the summer of 1940. Though there are routes to our ranch house that are now mostly paved, I still—unless it's raining—usually take the dirt road. After so much speeding across the country, once in a while it's nice to move slow.

I've always wanted to be on the go; and yet the agents of my mobility have never interested me very much. I was never romantic about, or very interested in, horses; all I expected of them was that they refrain from biting me or falling on me. The same holds for cars: I just want them to be heavy enough to carry me comfortably and I expect them to keep running until they get me where I am going. Otherwise I seem to lack entirely the mechanical dimension. I've never desired or yearned for a particular make of car; I have never fantasized about cars or invested any emotion in them. I just want a vehicle that won't lag and sputter if I want it to climb the Mogollon Rim.

I do love driving, though. In periods when I've lived in cities with good public transportation, where there's no need to drive— where, in fact, it's better not to drive—I soon began to feel choked and a little claustrophobic. I was allowed to drive on our ranch roads from about the age of nine; like everyone else in rural Texas then, I had a driver's license by the age of fourteen. When I was seven my parents moved from the ranch to Archer City, largely, I think, to spare me the grueling eight-mile-a-day school bus ride. This didn't mean that we stopped working on the ranch, it just meant we had to drive sixteen miles to get to it, and then sixteen miles back. In the fifties my father bought three sections of land at the opposite ends of the county; the two segments of the expanded ranch were twenty-five miles apart. For most of my childhood, adolescence, and young manhood my father and I, one or both, drove about one hundred miles a day, just going back and forth between the two halves of the ranch, doing the chores. Driving became as much a part of the rhythm of my existence as horseback riding had been for my grandfather. The winters were icy, the summer blazed, the work was hard—these short drives, as we moved from one part of the ranch to the other, often with two or three horses in the back of the pickup, provided little respites, interludes in which I could think, daydream, relax. Over the years what had begun as a necessary work habit became a need of a deeper sort. Being alone in a car is to be protected for a time from the pressures of day-to-day life; it's like being in one's own time machine, in which the mind can rove ahead to the future or scan the past. When I'm about to start a novel I've always found that driving across the country for a few hundred miles is a good way to get ready. I may not be forming scenes or thinking about characters—indeed, may not be thinking of much of anything on these drives. But I'm getting ready, all the same.

Though now, ambitiously, I might take all America as my province, the road that has meant most to me is the sixteen-mile stretch of dirt road that I drive from our ranch house to town. The road cuts through a number of ranches and just skirts the dairy farming country to the north. Though a very modest road, it has, over the years I've been driving it, produced a number of surprises. Once, when I was about ten, we were approaching the ranch after veering north to look at some pasturage when we saw a small barefoot boy racing along the hot road with terror in his face. My father just managed to stop him. Though incoherent with fear, the boy managed to inform us that his little brother had just drowned in the horse trough. My father grabbed the boy and we went racing up to the farmhouse, where the anguished mother, the drowned child in her arms, was sobbing, crying out in German, and rocking in a rocking chair. Fortunately the boy was not quite dead. My father managed to get him away from his mother long enough to stretch him out on the porch and squeeze the water out of him. In a while the boy began to belch dirty fluids and then to breathe again. The crisis past, we went on home. The grateful German mother brought my father jars of her best sauerkraut for many, many years.

Along that road there was for a time a one-room school, long abandoned, that was used as a voting place when election day came around. One day in the midst of the war I went horseback to that schoolhouse with my father so that he could cast what would be his final vote for President Roosevelt, whose rural electrification program had brought us the magic of electricity only a year or two before.

I don't think I quite comprehended what a president was, at that early juncture of my life, but I knew that President Roosevelt must be a very great man because, not long after the wires were strung along the dirt road to our house, we acquired a radio, whose heroes and heroines soon came to obsess me. I remember sitting sadly in the living room of our ranch house, with my parents and my grandmother, as the radio informed us that President Roosevelt was dead. The crackly reports continued for a long time—the president's body, it seemed, was being brought to Washington by special train; a reporter who seemed himself not very well composed informed us that thousands of Americans were lined up beside the railroad tracks, weeping as they waited for the president's body to pass.

The image of those thousands of weeping people, lined up by the railroad track, remained in my mind for many years.

Two years later, on that same dirt road, my father had a narrow escape. He had stopped on a little wooden bridge over a small creek to inspect the water gap beneath—I waited in the pickup, reading a comic book. My father found nothing amiss with the water gap and was about to get back in the pickup when I glanced at him and saw his face change. It was summer—my father was deeply tanned, but in an instant his face went white. Then I heard a faint buzz and saw my father whirl and grab a shovel out of the back of the pickup. The longest rattlesnake either of us had ever seen, a monster almost ten feet long, had struck at my father in silence, and just missed. The snake's head was almost touching my father's boot heel. My father killed the big snake with the shovel but, after that, had to sit on the running board of the pickup for almost half an hour, recovering himself. He was too shocked to speak. A neighboring rancher pulled up and stopped, examined the big dead snake, and helped my father pull his boots off, to be sure there were no fang marks anywhere. The rancher thought my father might be having a heart attack, but he wasn't. His color slowly came back, and we went on home.

I have ever since been extremely cautious when inspecting water gaps.

For most of their existence the dirt roads running from our ranch house to town had no names. They were just dirt roads, connecting with one another at obscure crossroads. If a stranger needed directions to a particular farm or ranch, he or she would be told to head on up such and such a dirt road, take a left or a right at the crossroads, and proceed until the correct house appeared. All these little county roads were more or less equal. They had been scrapped out by bulldozers and maintained by a number of county commissioners, whose road crews repaired erosion when it occurred and in general kept the roads serviceable.

Until the 1970s no one, so far as I am aware, gave any thought to naming these roads. After all, why name a dirt road?

Well, a reason loomed as 1980 approached, 1980 being the county's centennial year. In that year my neighbor Jack Loftin produced a compendious county history. Since the county was not populated by historically minded citizens, it was Jack, I suspect, who felt that the dignity of the centennial would be enhanced if the local roads were finally given names—even the dirt roads. This task was soon completed. It was not long before there were handsome green signs at every crossroads, with names on them. The dirt road that led to our ranch house was named Sam Cowan Road in its western reaches and Loftin Road as it passed Jack's house and proceeded east. Sam Cowan had been a pioneer rancher of great prominence.

The problem Jack ran into was that there had been more deserving pioneer families in the county than there were dirt roads. Some families got roads named for them, and some didn't, a circumstance that, for a time, produced wounded feelings. A few modest acts of protest, or even vandalism, occurred. Lifelong residents who didn't like the name given to the road past their homes would usually just cut down the sign. Some roads had to be named two or three times before the citizens who lived along them felt that Jack and his committee of road namers had finally got it right.

Driving east toward our ranch house on Sam Cowan Road

today, I frequently have the sense that I'm passing back through twelve decades of family history, to the time when my grandfather drove the same road in a wagon. The schoolhouse where everyone voted for President Roosevelt one last time is gone. The rancher who stopped to help my father recover from his scare with the big snake is gone. My father is gone, as are most of his brothers and sisters and even most of *their* children, my first cousins. The three cattlemen my father worked with most closely in the last two decades of his life are now very old men. Gordon Rucker is in his nineties—I often see him at the post office, getting his mail, a walker handy in the back of his pickup. Luke Smith and Howard Lyles are still active cowboys. Howard, in fact, now eighty-four, still occasionally ropes in rodeos off his twenty-seven-year-old horse. He is hoping some promoter will come up with an old-timer's event in which the combined age of horse and rider must exceed one hundred years.

Driving to the ranch house on Sam Cowan Road is to drive through a land of ghosts. I never fail to experience a pang of confusion and sadness as I pass the headquarters of the Ikard ranch. Bonny Ikard, who once lived there, was, for most of my father's life, his best friend. Then, a few years before my father died, he and Bonny had a falling-out, one so serious that they never made it up. I last saw Bonny Ikard at my father's funeral. He too died, soon after. I never quite understood what the quarrel was about, or why it cut so deep, but I never pass the Ikard ranch house without a sense of regret that these two men, neighbors and friends all their lives, had quarreled so seriously that they went to their graves without having made it up.

The country along Sam Cowan Road is unremarkable—pastureland, mostly, with cattle grazing here and there. There are a couple of ridges or rises where one can look northeast to Windthorst, but mostly it's just a road that cuts through undulating prairies where the mesquite is thickening yearly. On the Ikard lands some of the mesquite has been bulldozed, but not all. Sometimes I drive mechanically, thinking about something else. If I notice anything to admire on my drive it will usually just be the balletic swoop of a hawk.

I read the country casually, when I read it at all, in great contrast to my father, for whom land, grass, and sky composed a great, ever varying text whose interest he could never exhaust. What Proust is to me, the grasslands were to my father, a great subtle text which would repay endless study. He was a countryman, lifelong, and had a countryman's eye for the small variables of landscape that occurred even along a road he had driven many hundreds of times. Here a water gap might be out, a stretch of fence beginning to sag, a patch of grass ruined by a saltwater leak from some improperly tended oil well. When our well-to-do neighbor Mr. Bridwell let his beefmaster bulls into the bull pasture next to our land, my father would always slow when we approached a group of bulls. Sometimes, because of the magnificence of some of these great animals, he would come to full stop and just sit looking, arrested by admiration as a stroller in an art museum might be when brought face-to-face with a great picture—a Rembrandt, a Matisse.

I have looked at many places quickly—my father looked at one place deeply. Most of the citizens of Illiers (Combray) just saw the path that led to Swann's house as a path; it took Proust to see it as a world—which, on a homely scale, was how my father looked at

Sam Cowan Road or the other country roads he rode or drove along for some seventy years. For him those roads were crowded with memories. Now, as I age, I'm just beginning to understand how memory loops back on itself. Earlier memories advance, more recent ones recede. My short drives along Sam Cowan Road grow ever less simple, as people I met on it long ago crowd in again: the syphilitic, the old skunk woman, the iceman, the snake hunter who doubled as a bovine obstetrician, the cowboys, the Dutchmen, the Ikards, my grandparents, my father.

Every year the short drive to the ranch house on that dirt road becomes less of a simple thing.

SEPTEMBER

Seattle to Omaha via the 90, the 15, U.S. 87, U.S. 2, U.S. 3, U.S. 281, U.S. 50, and I-29

JUST BEFORE I LANDED in Seattle, Mount Rainier suddenly appeared out my window, as snowy as if there was no such thing as summer. There had been such a thing in Texas, though: the high was 112 the day I left.

Being inside the parking garage at the Seattle airport is rather like being inside a concrete tire: one circles around inside the tire a great many times before finally being allowed to come out.

Once out, though, I picked up my old friend the 90, last seen in Sioux Falls, South Dakota. It shoots east out of Seattle just a little south of Microsoft and soon slips through the Wenatchee Mountains, a part of the great Cascade Range. I had never driven in central or eastern Washington, and was not sure what sort of landscape to expect.

The writer a reader is apt to think about now, when visiting Seattle, is Raymond Carver, but when I first came there as a book scout in the early sixties, the writer I thought about was the wonderful and now little-acclaimed American poet Theodore Roethke. There is a fine little documentary about Roethke, called *In a Dark Time*. One thing he shared with Raymond Carver, and with the characters in Raymond Carver's stories, was depression. In both writers depression seems to be an inescapable part of life. Roethke fought it by taking baths, sometimes seven or eight a day. Though overshadowed in his lifetime by John Berryman and Robert Lowell, Theodore Roethke's work is here to stay. One poem still rings in my head whenever I think of him. It's called "Wish for a Young Wife":

> *My lizard, my lively writher,*
> *May your limbs never wither,*
> *May the eyes in your face*
> *Survive the green ice*
> *Of envy's green gaze;*
> *May you live out your life*
> *Without hate, without grief,*
> *And your hair ever blaze,*
> *In the sun, in the sun,*
> *When I am undone,*
> *When I am no one.*

Raymond Carver, like most masters, has been unfortunate in his imitators. The emotional algebra that seems subtle in his stories merely seems inarticulate in the stories of his imitators. Allowing for a huge difference in style, this is also true for Donald Barthelme, the other writer who has had a major effect on the contemporary short story. Barthelme was a literary acrobat who flew very high and, most of the time, managed to catch the bar. His imitators, like Carver's, plummet straight to earth.

Central Washington, when I descended to it from the Cascades, was an unexpected delight. It seemed to provide a kind of anthology of landscapes, some of them anticipations of country I would see farther east, or else recapitulations of country I had already seen, with pleasing alterations of plain and desert. The irrigated farms east of Moses Lake reminded me of Kansas or the Dakotas, but the occasional hummocks of golden grass were more like the grass in central California. The country bordering the Columbia River, as it carves out its great gorge, is rather Nevadalike. Some of it is wild horse country—an attractive line of wild horse sculptures set on a hill east of the Columbia. The river was very blue that day—I pulled off for a while just to watch it flow. I'm used to crossing the Columbia at Portland, where it's often hidden by mists or rain; encountered in the high desert, its beauty is more distinct. Rivers that flow through deserts often seem more beautiful, not to mention more welcome, than those that meander through the green parts of the world: there's the Nile, the Euphrates, the Colorado, the Rio Grande.

For most of the next two days I'm going to be driving not far from where Lewis and Clark walked, rode, or floated. What a relief it must have been for them to finally come to the Columbia. They had trudged over two passes, both formidable, and ridden a long way west with no certainty of success. They reached the navigable limits of the upper Missouri on August 17, 1805, and were over the Bitterroots by the Lolo Trail by September 22, a feat involving not only undaunted courage but extraordinary energies as well. In crossing the two passes they had the help of Sacagawea and her interpreter husband, Toussaint Charbonneau. Probably no more taxing or critical month occurred in the whole journey.

The Lolo Trail seems to call out the best in some people, but not necessarily in all people. The Nez Perce crossed it in 1877, while making their great flight to freedom. They got over the pass much quicker than their pursuer, one-armed General Howard, nicknamed General Day-and-a-Half because, throughout this long and dramatic pursuit, General Howard was always something like a day and a half behind the Indians. It's not likely that General Howard would ever have caught up. Unfortunately for Chief Joseph, Looking Glass, and the others, bad weather caught them and General Miles cut them off in the Bear Paw Mountains, only a day away from Canada.

The Lolo Pass is no cinch, even today. One of the pleasures of driving the 90 in this part of the world is that there are few trucks on it. The Bitterroot Range is steep, not an easy pull for the big rigs. The ascent is hard on their motors, the descent hard on their brakes. I only saw six, on this crossing.

I'm not entirely comfortable in Idaho—fortunately it's only seventy-five miles across the Idaho panhandle from Coeur d'Alene over the hills to Montana. I suppose my discomfort has to do with the Aryan Brotherhood and similar organizations, several of which make their official home in Idaho. In no state is there such obvious hatred of law and government—hard to explain, since there is scant evidence that there *is* much law and government in Idaho. A lot of frontier types who aren't quite up to Alaska hang out there, secure in the knowledge that they're in a part of the country where the outlaw mentality is still encouraged.

Ezra Pound was born in Hailey, Idaho. The Aryan Brotherhood has not thought to make him one of their heroes, though some of his more unfortunate views differed little from their own.

Ernest Hemingway, once Pound's good friend, chose to take his life at his home outside of Ketchum, Idaho. He had his own resentments of the U.S. government, particularly of the FBI, which he knew had been keeping an eye on him since the time of the Spanish Civil War. It was bird hunting that drew him to Idaho.

The writer most clearly identified with the state, though, was the prolific Vardis Fisher, born in Annis. Fisher was the mountain man of writers; from his mountain fastness he wrote some fifty books, including the massive WPA guide to the state. Vardis Fisher was not without talent, but none of his fifty books are quite good enough; his problem may have been an inability to slow down—his energy somehow overran his talent. Many of his books are interesting, but all of them feel hasty.

When I think of writings about Idaho, though, I forget Hemingway and Fisher and remember Clancy Sigal and that neglected American masterpiece *Going Away,* which contains a lovely ten pages about a visit the author paid to Coeur d'Alene just as he was starting to go away. Clancy Sigal gets just the right mixture of history and feeling into those pages, and I particularly like the modest way in which he describes an evening spent with a nice waitress named Kitty. What would America, much less Idaho, be without waitresses named Kitty?

Back in Washington State, before I started over the Lolo, I

came into wheat country, and I stayed in it for two days, across Washington, Montana, North Dakota. Except for one or two fields the wheat had already been cut and the combines had moved on north, into Alberta and Saskatchewan. Lots and lots of hay was being cut, though—in that country, winter could arrive any day.

As I sped down the eastern slopes of the Bitterroots I noticed water spurting into the air from two rather elegant yellow fountains in a glade by the roadside. A sign, briefly visible, merely says, *Elmer's Fountains.* It's a case of American creativity bursting out where one would least expect it, in the forests of the Bitterroots.

I spent the rest of a long, fine, late-summer afternoon drifting down through Montana, south to Missoula and then over to Helena. I only have to be in Montana—any part—about ten minutes to reconvince myself that it is easily the most beautiful American state. Many of its multitude of long, gentle glacial valleys have rivers running through them, in Norman Maclean's now famous phrase. The mountains are almost always in sight, but mainly as a shadowy blue backdrop to the beauty of the valleys and the plains. Most of the many small rivers—the Milk, the Marias, the Little Blackfoot, the Ruby, the Shields, the Sweet Grass—have small towns strung along them. The rivers, the valleys, the mountains, and the big sky manage, as nowhere else, to combine the grand vista with the intimate view, such as one might get along the Shenandoah or the Loire. On a fine day, with the great thunderheads rafting across the high sky, Montana can hold its own, for sheer visual beauty, with any landscape in the world.

I slipped right past Missoula, where there's usually a wild tangle of writers around the university—a kind of literary ghetto. On an earlier visit I once found myself in a jeep being driven by the estimable James Crumley. It was night, and we seemed to be proceeding straight up a quite steep mountain. When I asked James Crumley what we were doing he merely said that driving jeeps up cliffs was a time-honored local tradition.

It's just such time-honored traditions that keep me out of Montana, Wyoming, and Idaho for long stretches of time.

The next morning I picked up another old friend, I-15, last seen at Barstow, California, and rode it north to Great Falls, a town with so few exits that I had missed both of them before I realized I had already left town. From Helena up I had frequently been in sight of the Missouri, or if not, at least of the Missouri breaks, made famous in Tom McGuane's movie of that name. And I'm not far from Choteau, home of the late A. B. Guthrie Jr., whose novel *The Big Sky* created a kind of state phrase that even made it on license plates for a while.

Although both McGuane and Guthrie have written well of Montana, I'm driving north, above the Marias River, mainly to see the country where my favorite cowboy made his home. This was Teddy Blue Abbott, known simply as Teddy Blue, a nineteenth-century Kansas cowboy who made several trail-herding trips from south Texas into Montana and the Dakotas. With the help of a Montana newspaperwoman, Helena Huntington Smith, Teddy Blue produced what is in my view the single best memoir of the cowboy era. The book, *We Pointed Them North*, is as readable today as it was when it was published, sixty years ago. What distinguishes it is its vividness, its exuberance, and its candor. That there were whores in the west, and that he spent as much time with them as possible, is not a fact Teddy Blue bothered to ignore.

When the trail-driving days were over Teddy Blue married one of the half-Cree daughters of the famous—or in some quarters, infamous—pioneer cattleman Granville Stuart, and spent the rest of his life ranching along the Milk River, in north central Montana. I had never been to the Milk River and wanted to see where Teddy Blue chose to live when he finally left the long trail.

The reason Granville Stuart is infamous in some quarters is because of the severity of the justice meted out by the vigilantes who rode under his orders. In the 1880s the vigilante movement that he led resulted in the hanging of some thirty-five men—mainly for rustling in its various forms. The Montana ranchers were more effective than their fellow cattle barons in Wyoming. The latter produced the rather peculiar Johnson County War. Both struggles of large versus small interests in the west have often been treated in movies. The vigilantes in *The Missouri Breaks* would have been Granville Stuart's men; the Johnson County War was, of course, exhaustively treated in Michael Cimino's *Heaven's Gate*.

The ruthlessness and brutality that were common in the nineteenth-century west have generally been poeticized. It's been turned into pastoral, a transformation in which the landscape itself conspires. It's so beautiful that one's tendency is to forgive it, as a beautiful woman will be forgiven when a plain one won't. The lakes of blood that have soaked into all that fine soil are invisible now—and still the beauty remains.

At Havre, Montana, I turned east on U.S. 2 and was soon in Teddy Blue's home country. The narrow, green Milk River, the northernmost stream of any significance in that part of the west, is with me for almost three hundred miles, looping now north and now south of the highway. For much of that way the mighty Missouri itself is off my right elbow, its low bluffs always in sight.

Even the smallest Montana towns have something they call a casino now—even the humble convenience stores will have a long, dim room packed with game machines. All of the mini casinos are full. Truckers park their rigs along the curbs and hurry in to play a little keno.

East of Chinook my luck vis-à-vis speeding tickets finally ran out. I was clocked going eighty in a seventy. The irony, of which the nice officer who stopped me was painfully aware, was that until May 28, 1999, Montana had *no* daytime speed limit—one's speed was dependent upon temperament and the capabilities of one's vehicle, a freedom which finally produced an intolerable level of carnage on the highways. It may be that the officer just stopped me out of loneliness—his car and mine were the only cars in sight.

At some point in the afternoon, somewhere between Fort Peck in Montana and old Fort Union in North Dakota, near the spot where the Yellowstone River flows into the Missouri, I realized that I had found paradise. For connoisseurs of prairie travel, U.S. 2 is the perfect road—the road into the spacious heart of the plains. It runs from Spokane, Washington, straight across Idaho, Montana, North Dakota, and Minnesota, all the way to Lake Superior. I had passed its juncture with I-35 not ten minutes into the first of these drives, the one that began in Duluth.

If one's passion is high plains travel, U.S. 2 is as good as it gets. The hay fields were golden, the plowed land a rich brown, the Missouri bluffs bluish, the sky a deeper blue, the thunderheads a brilliant white, the hummocky, rolling rangeland a somber gray.

The colors, all subtle except the thunderheads, were constantly shifting and recombining as the clouds blocked and then released the strong sunlight. Once into North Dakota, tightly packed fields of tall sunflowers alternate with hay fields or the occasional patch of unharvested wheat.

At a stretch of road repair just inside North Dakota I had a human encounter, of an only slightly absurdist nature. As I approached the long stretch of repair a small wobbly man with a very red face crawled out of what had seemed to be an abandoned car—he sternly gestured for me to pull well off on the shoulder. This I did. The small man wobbled to my window and announced that he had yet to have breakfast—it was then nearly 5 P.M. He went back to his car, sat on the fender, and ate a prepackaged sandwich. Then he came back over and announced that the pilot car was due back in only fifteen minutes—there was so much heavy machinery, he said, that it would be suicide for me to attempt to get through without the pilot car.

The pilot car showed up right on time and had guided me only half a mile across the long miles of repair work when it was almost crushed itself by a hurtling dump truck whose driver didn't notice either the pilot car or me. The small wobbly man had certainly been right about the heavy machinery.

East of Minot I began to edge back into the land of the *mille lacs.* Ponds and small waterways were everywhere, some of them fringed with wild rice, all of them speckled with ducks—teal, mostly.

Throughout that afternoon and the next morning the realization slowly grew on me that I had accidentally found something I hadn't really expected to find: the dream road, the good-as-it-gets road, the ideal path into the heart of the great steppe. U.S. 2 had everything—the widest vistas, the greatest skies, and more history than any one traveler could possibly hope to exhaust: Lewis and Clark, the Missouri, the mountain men, the Cheyenne, the Sioux, Sitting Bull, the Yellowstone, Teddy Blue. Custer, had he survived, could have raced back down the Yellowstone and caught a steamboat home, to glory and his Libbie.

At Rugby, North Dakota—the geographical center of North America—I turned north to drive the northern end of the road of my boyhood, highway 281.

U.S. 281 collides with Canada a few miles north of the small town of Dunseith, North Dakota. Its last stretch borders the Turtle Mountain Reservation of the Ojibwa people, one of whose tribal chairs had been the novelist Louise Erdrich's grandfather.

When I turned around at the northern head of the 281 the view when I looked south was not very different from the view I have looking north from the porch of our ranch house, more than a thousand miles away, down the plain.

That inconsequential spot of the prairie must have been where I had been tending all this time, because by the time I was halfway back to U.S. 2, I realized I had lost my urge to drive. Though the country was still spacious and still beautiful, some impulse had been blunted. My intention, once I had seen the top of the 281, had been to go east to Grand Forks, on the Red River of the North. I had long been interested in the big trouble in that part of the country: the métis secessionist movement led by Louis Riel, who was hung in 1885. There is an excellent book about this interesting revolt: *Strange Empire,* by the now almost forgotten Montana historian Joseph Kinsey Howard. The nomadic peoples of the Red River of the North had more than a little difficulty accepting the fact that an international border had been created across lands where they had always migrated freely.

But arriving at the top of the 281 seemed to have muted my interest in métis secession. I drifted south, across the lands of the Spirit Lake Nation—that is, the Sioux—and was passing through the small town of Sheyenne, North Dakota, when I saw a comment that had been painted, by someone, on the wall of a building. NOTHING WAS EVER LOST THROUGH ENDURING LOVE OF NORTH DAKOTA, it read. There it was, American creativity bursting out again. I felt an immediate uplift—I don't know that I can claim enduring love for North Dakota, but I was certainly glad that I had come through Sheyenne and seen that sign.

A little later I stopped for breakfast in Carrington, a town evidently named for the rather ineffectual commander who, at Fort Phil Kearny, in 1866, allowed the impetuous Captain Fetterman and eighty men to ride out and get massacred by the waiting Sioux and Cheyenne. It was in this fight that Crazy Horse played the wounded bird and lured the eager soldiers to their doom.

Northern Montana and North Dakota are not heavily settled—people there might be excused for feeling themselves a little cut off from the world. What did they think of Kosovo, I wondered, or of W., as the columnists have begun to call George W. Bush, or of the senate race in New York State?

In the cafe the waitress lent me an already well used copy of the Fargo paper. A big headline said, GUILTY! A child murderer had just been convicted. Seven years before, a little girl had disappeared, presumably killed, her body never found. Possibly it had been weighted down and stuffed in the Sheyenne River.

There on the front page were her grieving parents, her weeping little sisters. Seven years had not dulled their pain—they were awash in what Auden called "the busy griefs." Inside was a picture of the victim, an exceptionally pretty, normal little midwestern girl whose neighbor just happened to be a sex offender. There was *his* picture too. His lawyer had tried to argue that the little girl had probably just run away, but the jury didn't buy that one for a minute. How much grief could anyone spare for Kosovo when there was that much grief right here at home?

A few miles farther on I happened to pass the Hamlin Garland Highway, which provided an irony almost as profound as my speeding ticket in Montana. The once famous author of *Main-Travelled Roads* now has a road named for him, but it definitely is not main, and it seems scarcely to be traveled at all.

At Jamestown, where I-94 crosses 281, I see a sign mentioning White Cloud, the white buffalo born a few years ago, to considerable rejoicing. There is a buffalo herd just south of Jamestown but none of the buffalo in it are white. Very likely White Cloud has been sequestered for his own protection. A white buffalo would be unlikely to survive two days along the shoulders of the 94. Somebody would probably haul back and shoot him.

South of Jamestown I began to feel a disinclination for sights, or even for food. U.S. 2 had turned out to be the best road, and the 281 had begun to seem, if anything, too long. I had seen a lot and no longer particularly wanted to look.

Long ago I habitually traveled with *Road Food*, a useful book by Jan and Michael Stern that locates good eats within a few miles of the interstates. On their advice I had once visited a cafe in Huron, South Dakota, where I ate the single best piece of pie I've ever eaten: sour cream and raisin. As I traveled south, Huron lay just a moment off my route, but I did not turn off to seek the ambrosial pie, nor did I journey over to Mitchell to revisit the Corn Palace, a rural performance center that had been the wonder of the prairies back around 1913. Then it had had the appearance of a great bulbous Russian church and was covered in real corncobs.

Gripped by a distinct, if mild, malaise I also passed up a chance to revisit Laura Ingalls Wilder's childhood home in nearby De Smet, where the author of *The Little House on the Prairie* settled with her family in 1879.

Some years ago I had a sobering realization about women, which was that there are just too many nice ones. One simply can't fall in love with, sleep with, or marry all the nice women— even serial marriers such as Mickey Rooney only manage eight or nine. One of the saddening facts of life is that there is always going to be a delightful woman somewhere who, for whatever accident of timing or attraction, simply slips by and recedes, to return only in dream.

As it is with women, so it is with roads. There are too many nice ones. I could go on for a long time, driving America's roads. I could see the sandhills of Nebraska, follow the old Oregon Trail along the North Platte, see the Tetons, dodge moose in Maine, slip down to Salt Lake City and remind myself what an inspired city planner Brigham Young had been.

But I can't drive all the roads. On even the narrowest highways that I've driven on these trips, and in even the smallest towns, there are signs pointing down even narrower highways to even smaller towns, many of which I will never see. There may be no mute inglorious Miltons in those towns, but there might be someone with simple good taste, like Elmer, who set those lovely fountains in the Bitterroots.

And there might be someone who, now and then, would feel like expressing a noble thought by painting it on a building: the conviction that nothing was ever lost through enduring love of North Dakota, for example.

When I was a boy, one of the first questions I asked my parents and grandparents was, where does the road go? I meant 281, of course, the road along which I was now hurrying home. Curiously, I was only interested in where it went to the north. I thought I already knew where it went to the south: Mineral Wells, Texas, a once-popular spa with two *grand luxe* hotels, where my panhandle uncles came every February with their well-upholstered wives to get out of the wind for a week or two.

When I asked my grandfather—the person assigned to deal with my questions when I was three and four—where the road went, he would merely allow that it went to Oklahoma. An old man then, near the end of his own road, he could see no reason why anyone would need to go farther north than Oklahoma.

In Teddy Blue's book *We Pointed Them North* there is an ignorant young cowboy who thought that north was a place, as Dodge City was a place. The other cowboys didn't disabuse him of this (to them) hilarious error. The ignorant cowboy believed that if one just kept going up the rivers, someday one would arrive at the place called North.

This morning outside of Dunseith, North Dakota, I *had* arrived at North—my personal north, at least. The arrival produced in me that empty feeling that I sometimes get after finishing a long work of fiction. For the moment, a question had been answered: so now what do I do?

Fiction, for all its subtlety and variety—a subtlety and variety more glorious even than the plains along U.S. 2—seems mainly to be asking two questions: Where does the road go? And how is one to marry?

Both in fiction and in person I'm still working on that second question, but I have finally been to where the road goes, and shouldn't need to go looking for a while.

Printed in the United States
By Bookmasters